Electricity and Magnetism in Biological Sys

Electricity and Magnetism in Biological Systems

D. T. EDMONDS

The Clarendon Laboratory
University of Oxford

OXFORD

UNIVERSITY PRESS

Great Clarendon Street, Oxford OX2 6DP

Oxford University Press is a department of the University of Oxford.
It furthers the University's objective of excellence in research, scholarship,
and education by publishing worldwide in

Oxford New York

Athens Auckland Bangkok Bogotá Buenos Aires Calcutta
Cape Town Chennai Dar es Salaam Delhi Florence Hong Kong Istanbul
Karachi Kuala Lumpur Madrid Melbourne Mexico City Mumbai
Nairobi Paris São Paulo Singapore Taipei Tokyo Toronto Warsaw

with associated companies in Berlin Ibadan

Oxford is a registered trade mark of Oxford University Press
in the UK and in certain other countries

Published in the United States
by Oxford University Press Inc., New York

A catalogue record for this book is available from the British Library

Library of Congress Cataloging in Publication Data

ISBN 0 19 850680 5 (Hbk)
ISBN 0 19 850679 1 (Pbk)

Typeset using the authors' LaTex files by Integra Software Services Pvt. Ltd.,
Pondicherry, India

Printed in Great Britain
on acid-free paper by
T.J. International Ltd,
Padstow, Cornwall

Preface

Our understanding of biological function is fast approaching the microscopic scale of atoms and molecules. To operate successfully on this scale it is necessary to adopt some of the techniques developed within the physical sciences for this purpose. In particular, it is known that outside the nucleus there are only two types of forces operating, electro-magnetic and gravitational. As gravity is far too weak on the molecular scale, it follows that only electromagnetic forces are significant. Thus on a molecular scale all biological function must eventually be understood in terms of electro-magnetic forces. This situation presents a challenge to biologists who have often had little training in this area and it presents an opportunity for physical scientists to contribute to an exciting and rapidly expanding field. This book aims to help biologists to master the theory of electricity and magnetism in a manner that is readily accessible to them and, in the later chapters, to point to some areas of biology within which these physical techniques may be applied directly.

The book has two distinct parts. Chapters 1 to 10 aim to present the theory of electricity and magnetism in a manner that is accessible to students of medicine, biology, biochemistry and chemistry. The approach is more descriptive and less reliant on algebraic manipulation than is the case in most physics textbooks. Whenever possible the proofs rely upon geometrical constructs and logical argument and do not require a deep mathematical knowledge. Despite this approach, the rigour of the treatment and the extent of the coverage are comparable to an undergraduate physics course. Within some early chapters are boxes which attempt to explain mathematical concepts that are essential, such as vectors, two-dimensional polar coordinates, and line, surface and volume integrals. Students with a stronger mathematical background can ignore these boxes without interrupting the logical flow of the treatment. No knowledge of vector algebra is required. Three appendices deal with (1) mathematical techniques, (2) the Boltzmann distribution and entropy and (3) an introduction to thermodynamics and the electrochemical potential. As many as possible of the topics discussed to illustrate the theory are directly relevant to biology. Conduction is discussed in terms of ionic diffusion in an aqueous solution rather than electronic conduction in a metal. Other examples include the treatment of the plasma membrane of an animal cell as a charge storage capacitor and the extent to which static and time-varying external electric and magnetic fields penetrate into a living cell. Chapter 7 gives practical advice about the properties of laboratory-generated magnetic fields suitable for applying to living cells.

At the end of each of the first ten chapters there is (1) a section describing in words the important topics covered in the chapter, (2) a listing of the most important equations that were derived and (3) a set of problems based upon the topics discussed in the chapter. Some of the problems are accompanied by worked solutions that expand the material dealt with in that chapter.

Chapters 11 to 16 apply the principles dealt with in the earlier chapters to selected topics where electricity or magnetism is known to play a direct role in the explanation of biological function. Within these chapters frequent reference is made to results derived in the earlier chapters. Chapters 11 to 16 can be used by biologically trained students as illustrations of the principles derived in Chapters 1 to 10 or they can be used by students trained in the physical sciences as an introduction to particular areas of biology where basic physical principles may be readily applied. One application chapters deals with the equilibrium properties of aqueous solutions, ionic hydration and the concepts of pH and pK_a and includes a discussion of how these parameters are changed by external electric fields and the dielectric environment. A second chapter discusses the changes in the properties of an ionic solution expected when it passes through a narrow pore such as a biological ion channel. Other chapters deal with the Debye layer, possible mechanisms for a magnetic animal compass and an introduction to the classical theory of nuclear magnetic resonance and magnetic resonance imaging. Chapter 15 describes an electrostatic model of a proton/ion coport or counterport which has many properties in common with the common but physically implausible mechanical models usually described in biochemical textbooks. The predicted properties of the model are explored by constructing a stochastic model of the device, based upon a Monte Carlo method, to illustrate the technique of computer simulation of biological function.

February 2001   D. T. E.
Oxford

Contents

PART I: THE BASIC THEORY

1. Electrostatic fields and voltage   3

1.1   Electricity, magnetism and biology   3
1.2   The sources of electrical forces   4
1.3   The electric field   5
1.4   Representation of electric fields   8
1.5   Gauss's law   9
1.6   Voltage   18
Topics covered in Chapter 1   23
Important equations of Chapter 1   24
Problems   25

2. How conductors shape an electric field   29

2.1   Electrical conductors   29
2.2   The electric field within a conductor   30
2.3   The electric field surrounding a conductor   33
2.4   The electrical capacitor   37
2.5   The animal cell plasma membrane as a capacitor   39
2.6   The capacitance of a single conductor   40
2.7   The spherical capacitor   41
Topics covered in Chapter 2   43
Important equations of Chapter 2   43
Problems   44

3. Ionic conductors   46

3.1   Ohm's law of conductivity   46
3.2   Ionic diffusion   48
3.3   Fick's first law of diffusion   51

3.4   Ionic motion in a constant electric field   53
3.5   Ionic motion with both an electric field and a concentration gradient   55
Topics covered in Chapter 3   56
Important equations of Chapter 3   57
Problems   59

**4.   Properties of the electric dipole and the application of
       Gauss's law when dielectrics are present**   61

4.1   The electric dipole   61
4.2   The interaction of electric dipoles with electric fields   66
4.3   Induced charges in dielectrics   68
4.4   Gauss's law with dielectrics present   72
4.5   The properties of the electric displacement $\tilde{D}$ 74
4.6   General procedure for solving problems in electrostatics   76
Topics covered in Chapter 4   77
Important equations of Chapter 4   78
Problems   79

**5.   The calculation of electric fields and voltages in the presence
       of dielectrics**   81

5.1   The dielectric constant of water   81
5.2   The electric field and voltage of a charged conducting sphere
      within dielectric material   82
5.3   Electric fields at the interface between two different dielectrics   84
5.4   Capacitors filled with dielectric material   86
5.5   The electrostatic self-energy of an isolated conductor   88
5.6   The extra electrostatic energy of an assembly of charges   91
5.7   The energy stored in an electric field   92
5.8   The cell plasma membrane as a barrier to ions   93
Topics covered in Chapter 5   94
Important equations of Chapter 5   95
Problems   96

6.   Static magnetic fields   100

6.1   The magnetic field and Gauss's law   100
6.2   The interaction of a magnetic dipole and a magnetic field   104
6.3   The magnetic field of a macroscopic current loop   106
6.4   The magnetostatic potential of a magnetic dipole   107
6.5   The magnetostatic potential near a current loop and Ampère's law   108

6.6   The differences between the electrostatic and magnetostatic potentials   112
Topics covered in Chapter 6   115
Important equations of Chapter 6   116
Problems   117

7.   The generation of magnetic fields   119

7.1   The magnetic field around a straight current-carrying wire   119
7.2   The long solenoid   121
7.3   The Biot–Savart Law   123
7.4   The single coil and the Helmholtz pair of coils   125
7.5   Practical coils for the generation of laboratory magnetic fields   127
Topics covered in Chapter 7   128
Important equations of Chapter 7   129
Problems   130

8.   Magnetic polarization of material   132

8.1   Magnetic material   132
8.2   Modification of Ampère's law by induced magnetic moments   133
8.3   Properties of the vector $\tilde{H}$ 138
8.4   Boundary conditions for $\tilde{B}$ and $\tilde{H}$ 139
Topics covered in Chapter 8   140
Important equations of Chapter 8   141
Problems   142

9.   Induced electric and magnetic fields   145

9.1   Faraday's law of induction   145
9.2   An application of Faraday's law   147
9.3   The screening of induced electric fields in biological tissue   149
9.4   The displacement current   150
9.5   Maxwell's equations   152
Topics covered in Chapter 9   153
Important equations of Chapter 9   154
Problems   155

10.   The motion of a charged particle in electric and magnetic fields and relativity   158

10.1   The Lorentz force   158
10.2   The motion of a free charged particle in a static magnetic field   159

10.3   The Larmor theorem   161
10.4   Diamagnetism   164
10.5   Special relativity and magnetism   166
Topics covered in Chapter 10   168
Important equations of Chapter 10   169
Problems   170

PART II: APPLICATIONS

11.   Ions in aqueous solution and the ionization of acids and bases   175

11.1   Ions in aqueous solution   175
11.2   The dissociation of the water molecule and pH   178
11.3   Ionizable residues   179
11.4   The effects of electric fields on the ionization of acid and
       basic residues   181
11.5   The effects of the electrical polarizability of the enviornment
       on the ionization of residues   182
References   185

12.   The Debye Layer   186

12.1   The basic electrostatics   186
12.2   The electric field and voltage at the surface for a given surface
       charge density   188
12.3   The variation of voltage with distance from the surface   191
12.4   The variation of ionic concentration with distance from the
       charged surface   194
12.5   How reliable is the simple theory of the Debye layer?   195
References   196

13.   The behaviour of ions in narrow pores   197

13.1   Ion channels in biology   197
13.2   The electrostatic self-energy of an ion in a narrow
       water-filled pore   198
13.3   Enhanced electrostatic interaction within narrow
       water-filled pores   201
13.4   Interactions between ions and ionizable residues in the pore wall   204
13.5   The possible ordering of the water structure within narrow pores   205
References   207

14.   Possible mechanisms for a magnetic animal compass   208

14.1   The magnetic field of the Earth   208
14.2   The animal compass   210
14.3   Magnetic induction   211
14.4   The magnetite compass   215
14.5   The free radical magnetic field detector   218
References   222

15.   An electrostatic model of a proton/ion or an ion/ion coport or counterport   223

15.1   The ionic coport and counterport   223
15.2   A simple mechanical model of a counterport   224
15.3   An electrostatic analogue of the mechanical model for a proton/ion counterport   225
15.4   Kinetics of the model   227
15.5   A Monte Carlo computer simulation   229
References   234

16.   An introduction to the semi-classical theory of pulsed nuclear magnetic resonance   235

16.1   Classical angular momentum and the Larmor theorem   235
16.2   The rotating frame   237
16.3   Application of a small-amplitude rotating magnetic field and magnetic resonance   239
16.4   The detection of nuclear magnetic resonance, the 90° pulse and the free precession signal   241
16.5   The 180° pulse and the spin echo   244
16.6   Nuclear magnetic resonance as a structural technique on a molecular scale   246
16.7   Nuclear magnetic resonance as a structural technique on a macroscopic scale   248
References   250

Appendix 1: Mathematics   251

A1.1   Cartesian and polar coordinates   251
A1.2   The work done by forces and couples   253
A1.3   Vectors   255
A1.4   Vector products   257

A1.5   Vector calculus   258
A1.6   Integrals   259
A1.7   Geometrical vector theorems   262

Appendix 2: The Boltzmann distribution, entropy and detailed balance   263

A2.1   Disorder and the number of available states   263
A2.2   The Boltzmann distribution   265
A2.3   Entropy   266
A2.4   Detailed balance   266
A2.5   An entropic force   268

Appendix 3: An introduction to thermodynamics and the chemical potential   270

A3.1   The first law   270
A3.2   The second law   272
A3.3   The Gibbs function   272
A3.4   Uses of the chemical potential   274
A3.5   The tension in an 'entropic chain'   277

Appendix 4: Hints for the solution of and numerical answers to the problems   279

Index   284

PART I: THE BASIC THEORY

1. Electrostatic fields and voltage

The simplest example of an electrical force is the Coulomb repulsion that is measured between two point charges of the same sign situated in a vacuum. In this chapter the concept of an electric field which arises from this example is introduced and Gauss's law, which governs the properties of the electric field, is derived. The voltage or electrostatic potential is discussed and its relationship with the electric field is described.

1.1 Electricity, magnetism and biology

Physicists have discovered that, outside of the nucleus, there are only two types of two-particle force, **electro-magnetic and gravitational**, and on the molecular scale gravitational forces are far too weak to play any role. Thus on the microscopic scale of molecules, atoms and ions, **all forces are either electric or magnetic**. Other forces, such as those which are loosely called 'chemical' or 'mechanical', have their origin in electro-magnetic effects. For example, the mechanical integrity of a solid object such as a book or a table is due to electrical forces, and a book placed on a table experiences an upthrust from the table equal to its weight because the outer electrons of the book's cover repel electrically those on the surface of the table. On the macroscopic scale of everyday life it would be unnecessarily complicated to express all forces in electro-magnetic form and inventing other types of force such as 'mechanical' or 'chemical' is a convenient shorthand to describe familiar situations. The collective behaviour of some many-bodied systems can be explained by the tendency of all systems to move to a state which maximizes the entropy of the system and the surroundings, as briefly discussed in Appendix 2. This is sometimes described as due to the action of an **entropic force** (see Appendices 2 and 3), but on the molecular scale the only physical forces acting to bring about the maximization of the entropy are in fact electro-magnetic.

Our understanding of biological processes is rapidly becoming more detailed. **When we come to understand biological function fully on a molecular scale, it must necessarily be in terms of electro-magnetic fields and forces**. In fact, with a few exceptions that we will discuss later, magnetism does not play an important role in biology and the fundamental forces at work are almost exclusively electric. It is the aim of the first ten chapters of this book to describe electro-magnetic theory in a manner that is more easily accessible to students of medicine, biology, biochemistry and chemistry than the formal mathematical approach of physics books. Emphasis is placed upon verbal description and geometric solutions rather than on algebra.

As far as is possible the theory we derive will be applied to situations that have some relevance to biology. However, in the first two chapters, where we will discuss the properties of static (time-invariant) electric fields in a vacuum and how the presence of conductors can change the shape of these fields, biological applications are clearly limited. In the third chapter, when we deal with the properties of conductors, we will discuss an ionic solution as the conductor most relevant to biology. In subsequent chapters we will point out any relevant applications to biology of the theory we derive. Chapters 11 to 16 discuss some biological phenomena that can be explained in terms of the electro-magnetic forces introduced in the first 10 chapters. These chapters can also serve as an introduction to some aspects of biology where students trained in the physical sciences can immediately contribute.

1.2   The sources of electrical forces

The sources of electrical forces are charges. Most material is electrically neutral and has no net charge. For example, a typical neutral atom has a nucleus containing a fixed number of positively charged protons and is surrounded by a cloud of the same number of electrons which each have a negative charge of exactly the same magnitude as that of a proton. However, if an electron is removed from the atom it becomes positively charged because of the single uncompensated proton, while the freed electron carries a negative charge of the same magnitude. Such charged atoms or particles are called **ions**. Charge is measured in units of coulombs, denoted by C, and the magnitude of both the positive charge on the proton and the negative charge on the electron is 1.6×10^{-19} C. This is the smallest charge possible in nature and all charges, positive or negative, are integral multiples of this elementary charge. Despite the small size of this charge, the electrical force between charges is so strong that the force exerted by a single electron or proton is often easily detected.

It is found experimentally that when one charge of $+Q_1$ coulombs and another of $+Q_2$ coulombs are located a distance apart of R metres in a vacuum, they repel each other with a repulsive force of $F(R)$ newtons, where $F(R)$ is given by the expression

$$F(R) = K \frac{Q_1 Q_2}{R^2}. \qquad [1.1]$$

K is a constant given by $1/(4\pi\varepsilon_0)$, where ε_0 is a universal constant equal to 8.854×10^{-12} which is called the **permittivity of free space**. The 4π is inserted into

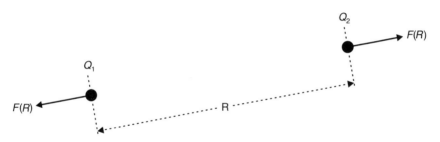

Fig. 1.1   The repulsive Coulomb force $F(R)$ between two positive charges Q_1 and Q_2 situated a distance R apart.

the definition of K at this stage in order that other formulae, derived from this equation, can be written in particularly simple forms. This expression for the repulsion between charges is called **Coulomb's law**. It can be seen that the repulsive force, which acts along the line connecting the two charges, is directly proportional to the product of the two charges and inversely proportional to the square of the distance R between them. The situation is illustrated in Fig. 1.1.

The electrical force, like gravity, is an example of 'action at a distance' such that one body influences another remote body even in a vacuum. We cannot 'explain' such forces in terms of models drawn from our everyday experience but are forced to accept them as the result of experiments. As described below, we deal with such effects by postulating that a charge creates an electric force field around it, even in a vacuum, and that this field interacts with the second charge to produce the force exerted between charges that is found experimentally.

Also like gravity, the electrical force is an example of an 'inverse square law', because of the dependence of the force $F(R)$ on $1/R^2$. If we remember that the force described by equation [1.1] is a repulsion, the equation correctly predicts algebraically that if the two charges have the same sign the force between them is repulsive, but if they have opposite signs (one positive and the other negative) the force between them is a negative repulsion or an attraction.

1.3   The electric field

Rather than invoking 'action at a distance' as an explanation of electrical forces, scientists have invented the concept of an electric field. The idea is that a force field called the electric field exists in the vicinity of any charge. The force of repulsion experienced by a second charge may then be explained as the force due to the interaction of this electrical force field with the second charge. In particular, it is postulated that around a single charge of $+Q_1$ coulomb in a vacuum there exists an electric field which is everywhere directed radially away from the central charge and has a strength $E_1(R)$ at a distance R metres away from the charge, where the magnitude of $E_1(R)$ is given by

$$E_1(R) = K\left(\frac{Q_1}{R^2}\right).$$ [1.2]

The constant K is the same as that in equation [1.1] above. The force $F(R)$ experienced by a second charge Q_2 placed in the field $E_1(R)$ is then defined to have an amplitude $Q_2 E_1(R)$, the product of the charge and the field strength, and it is defined to act in the direction of the field $\tilde{E}_1(R)$. The resulting equation for the amplitude of the force between the two charges $F(R)$ is given by

$$F(R) = Q_2 E_1(R) = K\left(\frac{Q_1 Q_2}{R^2}\right).$$ [1.3]

It is clear that this 'explanation' of the forces between charges given in equation [1.3] is entirely consistent with the experimental result given in equation [1.1].

Note that it is equally valid to say that the first charge Q_1 creates an electric field $E_1(R) = K(Q_1/R^2)$ at the second charge Q_2 which results in the repulsive force $F(R)$ given by equation [1.1], as it is to say that the second charge Q_2 creates an electric field $E_2(R) = K(Q_2/R^2)$ at the first charge Q_1 resulting in the same repulsive force.

From the first part of equation [1.3] it can be seen that **the strength and direction of the electric field $\tilde{E}$ at any point are defined as the strength and direction of the force $\tilde{F}$ acting upon on a charge of +1 coulomb placed at that point**. Electrical forces and fields are additive so that the total electric field at a given point due to many charges is simply the vectorial sum (see Box 1.1) at that point of the individual fields generated by the individual charges.

Box 1.1   Vectors

A vector represents a quantity such as a velocity or an electric field which has both an **amplitude** and a **direction**. A vector is printed in this book as a symbol such as $\tilde{v}$ which has a wavy line or tilde above it. The amplitude of this vector is a scalar number which may be printed as $|\tilde{v}|$, called the **modulus** of the vector $\tilde{v}$, or simply as the symbol v, without the tilde. The direction of this vector may be represented by a **unit vector** which has an amplitude of 1 but the direction of the original vector. The unit vector corresponding to $\tilde{v}$ is printed as $\hat{v}$ which consists of the symbol v with a caret sign (shallow inverted v) directly above it. Thus we may represent the vector as its scalar amplitude $|\tilde{v}|$ multiplied by the unit vector which indicates its direction

$$\tilde{v} = |\tilde{v}| \cdot \hat{v} = v \cdot \hat{v}.$$

A vector may also be represented by its **components**, which in three-dimensional Cartesian coordinates are (v_x, v_y, v_z). The three components are scalar numbers which represent the perpendicular projections of the vector on the three perpendicular Cartesian axes. Thus $v_x = v\cos(\theta_x)$, where θ_x is the angle between the direction of the vector $\tilde{v}$ and the x-axis, and similarly with v_y and v_z. We may represent the vector in terms of its components as in the equation

$$\tilde{v} = (v_x, v_y, v_z).$$

Applying Pythagoras's theorem in turn to the three components, we find a relation between the square of the amplitude of a vector and the sum of the squares of its Cartesian components:

$$v^2 = v_x^2 + v_y^2 + v_z^2.$$

The summation of the effects of two vectors acting at a given point must take into account both their amplitudes and their directions. It may be accomplished by the addition of their components as shown in Fig. B1.1. Here we sketch two vectors $\tilde{v}_1$ and $\tilde{v}_2$ in two dimensions such that the x- and y-components of the two vectors are (v_{x1}, v_{y1}) and (v_{x2}, v_{y2}), respectively. The sum of the two vectors is $\tilde{v}_3$ which is obtained by placing the two vectors $\tilde{v}_1$ and $\tilde{v}_2$ nose-to-tail as in the figure.

It can be seen that the x- and y-components of the vector sum $\tilde{v}_3$ become simply the sum of the x- and y-components of the individual vectors that are added so that

$$\tilde{v}_3 = (v_{x3}, v_{y3}) = (v_{x1} + v_{x2}, v_{y1} + v_{y2}).$$

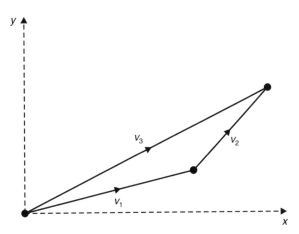

Fig. B1.1 The addition in two dimensions of two vectors $\tilde{v}_1$ and $\tilde{v}_2$ to yield the equivalent resultant vector $\tilde{v}_3$.

This process can be expanded to encompass the addition of any number of vectors in three dimensions. As an example we will calculation of the total electric field $\tilde{E}_{tot}(x, y, z)$ at the position (x, y, z) which is the resultant of many individual electric fields $\tilde{E}_i$ acting at the same point. First we find the components of the individual fields along the three Cartesian coordinate axes so that for each $\tilde{E}_i$ we determine $E_{x,i}$, $E_{y,i}$ and $E_{z,i}$. We then form the algebraic sum of the components of the individual vectors along each of the Cartesian axes to determine the components of the total vector $E_{x,tot}$, $E_{y,tot}$ and $E_{z,tot}$ along the same axes so that

$$E_{x,tot} = \sum_i E_{x,i}, \quad E_{y,tot} = \sum_i E_{y,i}, \quad E_{z,tot} = \sum_i E_{z,i}.$$

Now we have determined both the amplitude and the direction of the total electric field $\tilde{E}_{tot}$ in terms of its components along the Cartesian axes such that

$$\tilde{E}_{tot} = (E_{x,tot}, E_{y,tot}, E_{z,tot}).$$

See Appendix 1 for a more complete treatment of vectors.

1.4   Representation of electric fields

One way to represent electric fields in a diagram is to use **lines of force**. These lines are drawn everywhere in the direction of the electric field such that the density of the lines at any point is proportional to the strength of the field at that point. From this definition it may be deduced that lines of force start on positive charges and end on negative charges and may never cross each other. As an example, Fig. 1.2 shows the lines of force which represent the electric field surrounding a single isolated charge Q and which lie in a single plane.

The lines radiate outward from the centre of the charge in all directions in space with an equal density. They are most dense near the charge where the field is greatest (see equation [1.2]) and become further apart, and thus less dense, as we move away from the charge where the field grows weaker. Were we to draw two spheres of different radii centred on the charge, we would expect, if none of the lines of force terminate, that the total number of lines that cut each sphere is the same. The surface area of a sphere of radius R is $4\pi R^2$ so that the number of lines cutting a given surface area of a sphere must be inversely proportional to the square of the radius of the sphere and hence inversely proportional to the square of the distance from the charge. The fact that the experimentally measured strength of an electric field and the density of the lines of force, measured in this way, both obey inverse square laws for R is the justification of the assumed relationship between these two quantities.

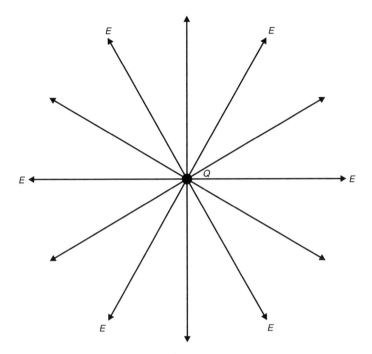

Fig. 1.2   A representation of the electric field $\tilde{E}$ in a single plane which contains a positive charge Q.

Lines of force are useful in diagrams illustrating the strength and direction of electric fields, but too close a dependence upon them can lead to problems. In the early days of electricity theory much reliance was placed upon lines of force and the question soon arose as to whether the lines of force, representing an electric field, were fixed relative to the charges creating the field or were fixed in space in an imaginary substance called the **ether** which was supposed to permeate all space, including a vacuum. In fact, in many situations, a logical development of a theory based upon lines of force would lead to the conclusion that lines of force must be fixed in the ether. Despite much experimental effort, no evidence for the presence of an ether has ever been found and, partly for this reason, we no longer rely upon the concept of lines of force for quantitative calculation. Today we rely directly upon the predictions of equations such as [1.2] and [1.3] without associating the electric fields they define too closely with lines of force.

1.5   Gauss's law

Gauss's law is the single most important law governing the properties of electric fields. It is a conservation law that was invented to deal with a flowing fluid such as water and is particularly easy to understand in that case. Consider a source of fluid which ejects a given mass S of incompressible fluid every second. The symbol S, defined in this way, represents the strength of the fluid source. It is self-evident that, in the steady state, the same mass of fluid will pass through a closed surface of any shape that encloses the fluid source provided that there exist no other sources or any sinks for the fluid within that surface. It is also evident that, for any closed surface that does not include the fluid source, the net flow of fluid into or out of the volume contained by that surface will be zero. Any flow of fluid into the surface will be exactly balanced by an equal outward flow in the steady state. The concept may be expanded to include both positive sources of strength S_i which **emit** a mass S_i of fluid every second and negative sources or sinks S_j which **absorb** a mass S_j of fluid each second. The strength of a sink is represented by a negative number S_j such that $S_j = -|S_j|$. Gauss's law in this case simply states that the total mass flow per second of fluid flowing outward across the surface of any closed volume is given by the algebraic sum of the strengths $\sum_i S_i$ of all the fluid sources and sinks contained within that volume.

Box 1.2   The Surface Integral

A **two-dimensional or surface integral** is used to calculate the area of a given surface or the flow or flux of a substance through that surface. To calculate the area of a surface, it is first divided into a large number of very small incremental areas of known size. The total area is then given by the sum of all these incremental areas required to cover the surface exactly once. Such a summation is represented by a double integral. As a particularly simple example we will compute the area of a plane rectangle of length a and height b. We choose two-dimensional Cartesian coordinates x and y such that the x-axis is parallel to the edge of the rectangle of length a and the y-axis is parallel to the edge of the rectangle of height b and such that the bottom left-hand corner of the rectangle is at the origin of coordinates. At the point $P(x, y)$ with coordinates x and y the increments in x and y are dx and dy (see Fig. B1.2a and the discussion of Cartesian coordinates in Appendix 1). These increments are very small, and in fact strictly speaking we take the limit as dx and dy tend to zero when using any increments in a calculation. We may divide the total area into small elemental areas $dS = dxdy$. The total area S is then given by the sum of the elemental areas dS such that the rectangle is covered exactly once. Such a summation is represented by a two-dimensional or surface integral when the range of the two variables x and y is specified. In this case $0 \leq x \leq a$ and $0 \leq y \leq b$ and the double integral is written as

$$S = \int_x \int_y dS = \int_{x=0}^{x=a} \int_{y=0}^{y=b} dxdy = \int_{x=0}^{x=a} dx \int_{y=0}^{y=b} dy = ab.$$

In this simple case the double integral may be written as the product of two independent single integrals each over a separate variable x or y.

As a more general example in two dimensions we will consider a plane area contained within the curve shown in Fig. B1.2b. We again use two-dimensional Cartesian coordinates as shown in the figure. The curve that encloses the required area is defined by an equation of the form $F(x, y) = 0$ so that for each value of the variable x the corresponding values of y that lie on the curve may be determined by solving the equation for that value of x. Similarly, the values of x on the curve that correspond to a particular value of y may be obtained.

Once again, at the point $P(x, y)$, the increments in x and y are dx and dy and the elemental area is $dS = dxdy$ which is shaded darkly in Fig. B1.2b. To find the total area within the curve we need to sum all such small areas dS so that the area is covered exactly once. As with the rectangle, we could in principle represent this sum by a double integral with the limits on x and y such that the area covered is within the given curve:

$$S = \int_x \int_y dS = \int_x \int_y dxdy.$$

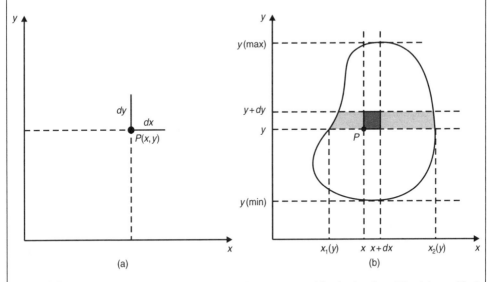

Fig. B1.2 (a) Two-dimensional Cartesian coordinates. Any position in the plane $P(x, y)$ is specified by giving the distances x and y along the horizontal and vertical axes respectively. The increment in area is the product $dxdy$ of the two increments in length. (b) The calculation by integration of the area of a plane figure. The area of the lightly shaded strip at the coordinate y is calculated as the integrated length $L_1 = \int_{x_1(y)}^{x_2(y)} dx$, multiplied by dy. The total area of the figure S is then given by integration over y so that $S = \int_{y(min)}^{y(max)} L_1 dy$.

In practice it is necessary to perform the summation over each of the two coordinates x and y in turn. Thus the lightly shaded strip in Fig. B1.2b has a height of dy and a length obtained by a summation of dx between the limits $x = x_1(y)$ and $x = x_2(y)$ where these limits are determined by solving the equation of the containing curve $F(x, y) = 0$ for the relevant value of y. If the area of the strip is given by $\Delta S(y)$ it may be written in terms of an integral over x as

$$\Delta S(y) = dy \left(\int_{x=x_1(y)}^{x=x_2(y)} dx \right) = dy[x_2(y) - x_1(y)].$$

Because dx and dy are so small, a negligible error is introduced because the left and right limits of the area $\Delta S(y)$ are not vertical. The total value of the area S within the curve is then obtained by summing all possible areas $\Delta S(y)$ between the extreme values of y which in the figure are $y = y(\text{min})$ and $y = y(\text{max})$, such that the total area within the given curve is covered exactly once. Such a sum is represented by the double integral

$$S = \int_{y=y(\text{min})}^{y=y(\text{max})} \left(\int_{x=x_1(y)}^{x=x_2(y)} dx \right) dy = \int_{y=y(\text{min})}^{y=y(\text{max})} [x_2(y) - x_1(y)]dy.$$

Note that in this case the two integrals are not independent as the integral over the variable x depends upon the value of y.

In principle, the surface area S of a three-dimensional surface defined by an equation of the type $F(x, y, z) = 0$ can be found in the same manner by summing the small elemental areas dS that exactly cover its surface once. Although the calculation may be difficult in practice, such a surface area is simply represented by the integral

$$S = \iint_S dS.$$

As well as measuring an area, a double or surface integral may measure J, the **flux of a quantity passing through a given surface**. This flux is defined as the total 'amount' of that quantity passing through the surface each second. We will take the example of the volume of an incompressible fluid that passes outward through the given surface S each second. At a particular point P on the surface let the velocity of the fluid be represented by the vector $\tilde{v}$. At P we define a very small element of the surface with an area dS and then calculate the component of the velocity v_N that points outward normal (perpendicular) to the surface element dS. We can also calculate the two components v_{T1} and v_{T2} of the velocity $\tilde{v}$ at P which are perpendicular to the direction of v_N. Because v_{T1} and v_{T2} are perpendicular to v_N they are tangential to the small area dS and do not penetrate through it. The small volume of the fluid that passes through dF is given simply by $v_N dS$. To calculate the total volume of fluid J (the fluid volume flux) that flows outward through the total surface each second we must add the increments dF such that the areas dS exactly cover the complete surface once. We can represent this as a surface integral

$$J = \iint_S dJ = \iint_S v_N dS.$$

In electricity theory we define **the flux of the electric field $\tilde{E}$** through a small area dS at a point P on a given surface as $dJ = E_N dS$, where E_N is the component of the field at P which is normal to the surface element dS. The electric field has replaced the velocity of the flowing fluid. Once again the two components of $\tilde{E}$ that are perpendicular to E_N do not contribute to the flux through dS. The total flux of the electric field that passes outward through the whole surface S is the sum of such elements of flux and it may be represented by the integral

$$J = \iint_S dJ = \iint_S E_N dS.$$

The summation represented by the double integral must be such that the elements of surface area dS cover the total surface area exactly once.

See Appendix 1 for a fuller treatment

To apply this law to electricity we must invent a property of the electric field that corresponds to the flow of the fluid. We define a quantity J_{tot} to represent the **total flux of the electric field** which passes outward through a given closed surface. At a given point on the surface we define a small area dS, centred on the chosen point, which is penetrated by the local electric field $\tilde{E}$ that acts at that point. The elemental flux dJ of the field through this element of surface is then defined as the product of E_N and dS, $dJ = E_N dS$, where E_N is the component of $\tilde{E}$ perpendicular to the surface element in the direction of the outward-pointing normal to that surface at the chosen point. The other two components of the field, which contribute to E_T, perpendicular to the normal, are tangential to the surface element and do not contribute to the flux of the field passing through this surface element.

The situation is illustrated in Fig. 1.3 which is taken to be in the plane of the electric field $\tilde{E}$ at the chosen point on the surface. The component of the electric field

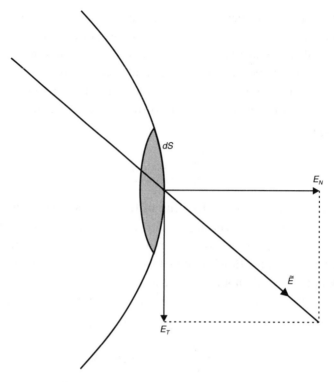

Fig. 1.3   An electric field $\tilde{E}$ penetrating through the centre of a small area dS on a curved surface. E_N and E_T are the components of the field normal and tangential to the surface at dS. The diagram is in the plane that contains the electric field.

E_N perpendicular to the surface is shown, as is E_T which represents the sum of the components of $\tilde{E}$ perpendicular to E_N. The total flux of the electric field out of a closed surface of any shape, which we represent by the symbol J_{tot}, may be written mathematically as a surface integral (see Box 1.2). The surface area of the closed surface is first divided into a large number of very small elementary areas dS. At each of these areas the outward-pointing normal component of the electric field E_N is calculated. The surface integral in the equation for the total flux of the electric field written below represents the sum of all the products $E_N dS$ summed over all the elementary surface areas dS of which the closed surface is composed:

$$J_{\text{tot}} = \iint_S dJ = \iint_S E_N dS. \qquad [1.4]$$

Let us now calculate the total flux $J_{\text{tot}}(Q)$ of electric field passing through the surface of the sphere of radius R which surrounds a single charge Q placed at its centre.

This is particularly simple to calculate as the electric field is constant and equal to $K(Q/R^2)$ everywhere on the spherical surface as given by equation [1.2]. The field, radiating outward from the single charge is also always in the direction of the outward-facing normal to the spherical surface, so that $E_N = E(R)$. The total outward flux of $\tilde{E}(R)$ generated by this charge Q is then given by

$$J_{tot}(Q) = (\text{Area of sphere}) \times (\text{Amplitude of } E)$$

$$= 4\pi R^2 \left(\frac{Q}{4\pi\varepsilon_0 R^2} \right)$$

$$= \frac{Q}{\varepsilon_0}. \tag{1.5}$$

Note that the total outward flux depends upon the magnitude of the charge Q but does not depend upon the radius of the sphere. Thus the total outward flux must be the same for all spheres centred on the charge Q. This is a direct consequence of the law, given in equation [1.2], which shows that the electric field of a charge at radius R is proportional to $1/R^2$ and the fact that the surface area of spheres increases with R as R^2. If the charge we are considering is negative such that $Q = -|Q|$ then there is a net **inward** flux of the electric field so that the **outward** flux given by Equation [1.5] is negative but is still given by $J_{tot}(Q) = -|Q|/\varepsilon_0$.

It can easily be proved, using some equations in solid geometry, that the total outward flux of the electric field as defined in equation [1.4] through a closed surface of **any shape** surrounding a charge Q is also given by Q/ε_0. In this electrical form of Gauss's law, the outward flux of the electric field has replaced the outward fluid mass flow per second, and the quantity Q/ε_0 has replaced the strength S of the fluid source. As the sign of the electric field depends linearly on the sign of the charge which gives rise to it, as in equation [1.2], the total flux of many charges, positive and negative, contained within a closed surface is just the algebraic sum of the fluxes of the individual charges. **Gauss's law then states that the flux of the electric field outward through any closed surface is given by the total algebraic sum of the charges $Q_{tot} = \Sigma_i Q_i$ contained within that surface, divided by ε_0.** Mathematically, using equation [1.4], this law is written as

$$J_{tot}(Q_{tot}) = \iint_S E_N dS = \frac{Q_{tot}}{\varepsilon_0}. \tag{1.6}$$

In this expression Q_{tot} is the total algebraic sum of all the charges contained within that surface. As with the fluid, charges that lie outside the surface contribute nothing to the total outward electric field flux through that surface.

Box 1.3   The Volume Integral

Just as a given area may be calculated by the summation of the incremental areas that compose the given area in a two-dimensional or surface integral, a given volume may be calculated by the summation of the incremental volumes that compose the given volume using a **three-dimensional or volume integral**. Thus the volume integral is a summation over three dimensions that can measure a volume or the quantity of some substance contained within a specified volume. In three-dimensional Cartesian coordinates, the element of volume at the position (x, y, z) is $dV = dxdydz$, where dx, dy and dz are the increments in x, y and z. The volume V of specified shape may be written as a triple integral

$$V = \int_x \int_y \int_z dxdydz = \iiint_V dV.$$

In the first integral the summation is explicitly shown over the three Cartesian coordinates which span particular ranges such that the specified volume is covered exactly once by the elements of volume $dxdydz$. The second integral is more general and can represent the summation of the elementary volumes dV in any coordinate system such that the total specified volume is covered exactly once by the summation over the elementary volumes dV.

The total amount of some quantity within a given volume may also be calculated using a volume integral if the density or amount of the quantity per unit volume is known throughout the volume. As an example, the total mass M contained within a specified volume may be determined if the density or mass per unit volume at the position (x, y, z) is known to be $\rho(x, y, z)$. The mass contained within the element of volume $dxdydz$ then becomes $\rho(x, y, z)dxdydz$ and the total mass within the specified volume may be written as

$$M = \int_x \int_y \int_z \rho(x, y, z)dxdydz = \iiint_V \rho(x, y, z)dV.$$

Again the summation is such that the total specified volume is covered exactly once by a summation over the elementary volumes dV.

See Appendix 1 for a fuller treatment.

For a charge distribution that is smeared over space rather than being concentrated into discrete charges, Gauss's law may be written as

$$\iint_S E_N dS = \frac{Q_{tot}}{\varepsilon_0} = \iiint_V \frac{\rho(x, y, z)}{\varepsilon_0} dxdydz. \tag{1.7}$$

The volume integral (see Box 1.3) calculates the total charge within the volume enclosed by the Gauss law surface, by dividing that volume into a very large number of very small elementary volumes $dxdydz$. Within the volume $dxdydz$ the total charge is given by the charge density or charge per unit volume at that position $\rho(x, y, z)$, multiplied by that elementary volume $dxdydz$. The volume integral over the coordinates x, y and z represents a sum of the products $\rho(x, y, z)dxdydz$ such that the elementary volumes $dxdydz$ compose the total volume within the closed surface. Gauss's law, as written in equation [1.7], then states as before that the flux of the electric field out of the closed surface is equal to the total charge contained within that closed surface, divided by ε_0.

It is clear from our discussion above that led to equation [1.5] that, because the surface area of a sphere of radius R increases as R^2, only fields which obey an inverse square law will obey Gauss's law. However, an important extra condition for Gauss's law to be obeyed is that **the only sources and sinks of electric field are charges**. Our derivation of Gauss's law for a sphere or any other shaped surface relied upon the continuity of the electrical field in regions that do not contain charges. Only by changing the quantity of charge **within** a closed surface can we alter the total outward flux of electric field through that surface.

As a particularly simple application we consider two charges, a positive charge $+Q_1$ and a negative charge $-Q_2$, such that the magnitude of Q_1 is greater than the magnitude of Q_2. If we apply Gauss's law to small spheres drawn closely around each of the two charges we will deduce that the total flux of the electric field emanating from Q_1 is greater than that converging upon Q_2. Thus we can deduce using Gauss's law that some of the field originating at Q_1 will terminate on Q_2 but some cannot and must terminate on negative charges elsewhere.

We will apply Gauss's law often when deriving the strengths and directions of electrical fields. It is clearly important that the shape of the Gauss law surface is chosen to match the symmetry of the problem studied. When exploring the electric field around a charge or a spherical charged conductor it is sensible to use a spherical Gauss law surface. Then the electric field is, by the spherical symmetry of the charge distribution, constant over the spherical Gauss law surface and also everywhere perpendicular to that surface. In this case the flux of the electric field through the Gauss law surface is simply the surface area multiplied by the amplitude of the electric field at the surface, as in the example of the single charge discussed above. Similarly, cylindrical Gauss law surfaces are used when the charge distribution has cylindrical symmetry. Another commonly used Gauss law surface is the 'pillbox' a circular cylinder with plane ends perpendicular to the cylinder axis. This is often used when the electric field all points along the axis of the cylinder. Because the electric field is tangential to the curved surfaces of the cylinder there is no flux through them but only through the plane ends of the cylinder. Such a pillbox is used to study the change in the electric field as it passes perpendicularly through the plane interface between two different materials. It is possible to use the pillbox even when the field does not point along the cylinder axis by making the cylinder so short that the area of the curved surface is negligible in comparison with the area of its flat ends. The flux of the electric field through the curved surface can then be made negligible in comparison with the flux through the flat ends.

1.6   Voltage

Most people are familiar with the notion of voltage as some measure of the 'strength' of a battery, so that a common acid flash-light cell is labelled as producing 1.5 volts. Voltage is an important concept in the development of electrostatic theory and we will deal with it at some length. Voltage multiplied by charge is in fact a measure of energy, and a voltage difference is defined by the following statement. **The voltage rise dV between two positions close to each other in space is the energy in joules that must be expended (external work done on the electrical system) to move a unit charge (1 coulomb) between those two points.** The work done is proportional to the charge moved, so that to move a charge of q coulombs between two closely spaced points when the voltage rises by dV requires external work $+qdV$.

We will consider the case of a single charge of q coulombs in free space at a point where there exists an electric field E_X acting in the x-direction. To prevent the charge moving we must apply a force on the charge from an external agency, exactly equal and opposite to the force qE_X exerted on the charge q by the electric field. In this situation, with no dissipative frictional or viscous forces acting, the total energy of the system is conserved. Energy lost by the electrical system is gained by the external agency and vice versa. If we allow the charge to move an infinitesimal distance dx in the x-direction, the work done by the electrical system, defined as the force acting multiplied by the distance travelled by the point of application of the force in the direction of the force (see Appendix 1), is equal to $dW = qE_X dx$. The energy stored in the electrical system must drop by dW because of the work done, and thus the energy stored in the external agency rises by dW. In a formal sense we may say that the work **done on** the external agency when the charge moves is $+dW$ so that the work **done by** the external agency in the move is $-dW$. Applying our definition of voltage difference to this situation, we may deduce that

$$qdV = -dW = -qE_X dx, \quad \text{or} \quad dV = -E_X dx. \qquad [1.8]$$

The negative sign in this relationship is important. In words, it simply means that moving in the direction of a prevailing electric field necessarily involves a drop in the voltage experienced.

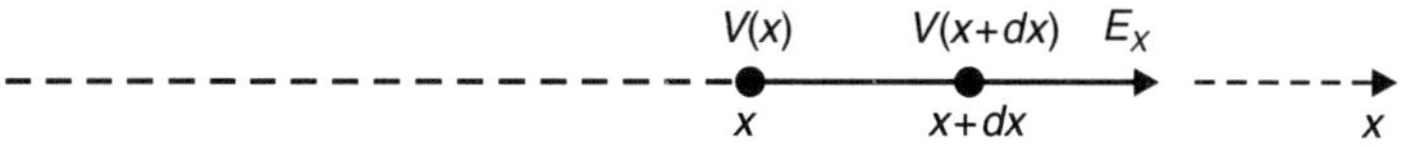

Fig. 1.4   The *x*-component of the electric field E_x passing through two points with *x*-coordinates x and $x + dx$. The voltage at these points is shown as $V(x)$ and $V(x + dx)$ respectively.

Dividing both sides of equation [1.8] by dx, we obtain the important relation

$$E_X = -\frac{dV}{dx}. \qquad [1.9]$$

As dx becomes very small, the ratio dV/dx tends toward a differential or gradient, so that **the electric field in a given direction is everywhere given by the negative gradient of the voltage in the same direction**. In three dimensions we have, by similar arguments, the relationships

$$E_X = -\frac{dV}{dx}, \quad E_Y = -\frac{dV}{dy}, \quad E_Z = -\frac{dV}{dz}. \qquad [1.10]$$

It is because of the relationships described in equation [1.10] that the units volts per metre ($V\,m^{-1}$) are used for the electric field.

As mentioned above, it is important to notice the negative signs in equations [1.8], [1.9] and [1.10] and to understand clearly their origin. This is shown in Fig. 1.4 which illustrates the drop in voltage between two points x and $x + dx$ located along the direction of the prevailing electric field E_x. If a positive charge at the point with coordinate x is to move to a point with coordinate $x + dx$ spontaneously under the influence of electrical forces, the energy stored in the electrical system must fall during the move so that $dV = V(x + dx) - V(x)$ is negative. Turning our attention to the prevailing electric field E_X in the direction of the move, this must be positive if a spontaneous move of the charge in this direction can occur because the force on a positive charge is in the direction of that field. Thus positive electric fields in a given direction are associated with a drop in the voltage in the same direction.

Box 1.4   The Line Integral

The line integral is a summation in one dimension. It may be used to calculate the length of a line or the change in some quantity along a specific path. We will consider a curved line in three dimensions that starts at the point a with Cartesian coordinates (x_1, y_1, z_1) and ends at the point b with coordinates (x_2, y_2, z_2). At a particular point (x, y, z) on the specified line, let the next increment in distance

along the line be represented by dR. Then the length of the line between its ends may be represented by L in the equation

$$L = \int_a^b dR.$$

The integral represents the summation of all the elementary displacements dR along the line such that the line is traversed exactly once between its starting point at a and its end point at b. If the summation is around a closed loop the integration symbol includes a circle:

$$L = \oint dR.$$

As an example let us now calculate the change in height h as we traverse a particular path on the surface of a hill between points a and b as defined above. At a particular point (x, y, z) on the path let the height be $h(x, y, z)$ and the next small element of distance along the path be dR. If the change in h as we move a distance dR along the path is given by dh then we may write an equation $dh = [dh(x, y, z)/dR]dR$, where $dh(x, y, z)/dR$ is the gradient of the height with distance along the direction of the next elementary step dR. The total increase in height, Δh, as the complete path is traversed may then be calculated using a line integral as below.

$$\Delta h = h(b) - h(a) = \int_a^b dh(x, y, z) = \int_a^b \left[\frac{dh(x, y, z)}{dR}\right] dR$$

In order to carry out the calculation we need to know $h(x, y, z)$ and hence $dh(x, y, z)/dR$ for all points on the surface of the hill along the chosen path.

In the case of an electric field $\tilde{E}$, the component of the field E_R resolved along the direction of the path increment dR at the point on the path (x, y, z), is given by equation [1.9] as $E_R = -dV(x, y, z)/dR$. Thus the total increase in the voltage V as we move along a particular path between the points a and b may be calculated using a line integral:

$$V(b) - V(a) = \int_a^b dV(x, y, z) = \int_a^b \left[\frac{dV(x, y, z)}{dR}\right] dR = -\int_a^b E_R dR.$$

So far we have dealt with voltage changes only over infinitesimal distances such as dx. To calculate the voltage change over larger distances along a particular path, we may divide the path into infinitesimal steps and sum the changes in voltage that result from each step along the path. We define the length of an infinitesimal step along a particular path as dR. The voltage change while performing this step is given by

$dV = (dV/dR)dR = -E_RdR$, where E_R is the component of the total electric field at that point in the direction of the dR (see equation [1.9]). Mathematically such a sum of infinitesimal steps is represented by a line integral (see Box 1.4) so that we may write the total integrated increase in voltage along a particular path between the points a and b as $V(b) - V(a)$, where

$$V(b) - V(a) = \int_a^b dV = \int_a^b \left(\frac{dV}{dR}\right)dR = -\int_a^b E_RdR. \qquad [1.11]$$

The integral is performed along the chosen path between the points a and b.

If the electric field $\tilde{E}$ is due to a single charge it can easily be proved geometrically that the line integral $\int_a^b E_RdR$ evaluated between any two points a and b as in equation [1.11] does not depend upon the path taken between a and b. This is a direct consequence of the fact that the electric field points everywhere radially away from the charge. However, any static electric field due to any distribution of charges is simply a linear combination of the individual fields due to the individual charges, so that we deduce in general that the line integral in equation [1.11], for any electric field created by static charges, **does not depend upon the path between the points a and b over which the integral is taken**. A very important consequence of this conclusion is that the value of the voltage change calculated along any path between a and b is the same. Thus, if we assign a value of voltage to a single point in space, the value of the voltage at every other point in space is uniquely determined. We may then say that **the voltage is a single-valued function of the position**.

This property of the voltage created by static electric fields is often illustrated by taking the line integral around a closed path starting and ending at the same point a. The line integral is necessarily zero as $V(a) - V(a) = 0$. This is often expressed in the form

$$\oint E_RdR = 0, \qquad [1.12]$$

where the integral sign with the circle indicates integration around any closed path. This relation is described in words by saying that **the line integral of the electric field around any closed path is zero**.

Clearly the reason why this relation holds is that along some parts of the path the component of the field along the path E_R is positive (voltage falling) and at others it is negative (voltage rising) and the total line integral around a closed path sums to zero. In fact, **equation [1.12] and Gauss's law are sufficient for the description of all the properties of static electric fields generated by fixed charges in a vacuum**. We will find later in Chapter 9 that other types of electric field may be generated by

Fig. 1.5 The electric field $E(R)$ at a point a distance R from a charge $+Q$.

time-varying magnetic fields and that these electric fields do not obey equation [1.12] and thus may not be described by a single-valued voltage function.

In our original definition of voltage we defined only a change in voltage and not an absolute value of the voltage. By convention the voltage very far away from any charges is taken as zero, so that a single absolute voltage may be assigned to any position in space by using the following definition. **The voltage at any point in space is the work done by an external agency when moving a unit charge (1 coulomb) from infinity to that point.**

We can use this definition to calculate the voltage distribution around a single charge Q. The whole system has spherical symmetry so that we assume that the voltage also has spherical symmetry. We will work out the voltage change by computing the line integral from infinity along a straight line directed towards the charge. When, as illustrated in Fig. 1.5, we are a distance R from the charge, the electric field is $E(R) = Q/(4\pi\varepsilon_0 R^2)$ directed diametrically away from the charge along the direction of dR, the increment in R. Thus the work done **by the electrical system** on a unit charge when the charge moves radially outward from a position given by R to infinity is given by ΔW_E in the equation

$$\Delta W_E = \int_{R=R}^{R=\infty} E_R dR = \int_{R=R}^{R=\infty} \left(\frac{Q}{4\pi\varepsilon_0 R^2} \right) dR = +\frac{Q}{4\pi\varepsilon_0 R}. \qquad [1.13]$$

But in a conservative system this is just the work done **by an external agency** to move the unit charge in the opposite direction from infinity to the position defined by R. Thus, from our definition of voltage change we may write

$$V(R) = V(R) - V(\infty) = \frac{+Q}{4\pi\varepsilon_0 R}. \qquad [1.14]$$

When deriving this result we have used our definition that $V(\infty) = 0$. Thus the distribution of voltage around a single charge Q is spherically symmetric and is directly proportional to the amplitude of the charge and inversely proportional to the distance from the central charge. Note again that the voltage drops when moving in the direction of the local electric field.

The voltage, also called the **electrostatic potential**, is a scalar quantity. It is a continuous function which covers all space and is represented at any point by one simple number. However, by differentiating this function with respect to x, y or z, as in equation [1.10], we may deduce the three vector components of the electric field at that point. Thus we may totally specify all three components of the electric field at any point by simply giving the single value of the voltage at every point. This is one important use of the concept of voltage.

The meaning of an electrostatic potential or voltage is not simple and requires some thought to understand fully. For example, it is not the same as the potential energy of the electric system. In some senses, voltage does not exist outside our heads. In the real world we can detect electric forces and hence electric fields directly using test charges, but we cannot directly detect voltage. Scientists have invented voltage as an intellectual concept because it is a simple scalar number which helps us to visualize what is happening in an electric environment. The essential idea is the relationship between the negative gradient of the voltage and the electric field expressed in equation [1.10].

A useful and familiar analogy to the relationship between the electrical field component and the gradient of the electrostatic potential or voltage is that between the force due to gravity and the gravitational potential energy. For a particle of mass m moving on a smooth surface of varying height h, the gravitational potential is mgh, where g is the acceleration due to gravity. Again the gravitational potential is an intellectual concept, while the forces acting on the particle are real and directly measurable. As the particle moves on the surface, the gravitational force in any direction at any position on the surface is given by the negative gradient of the gravitational potential along the surface in that direction. The maximum gravitational force along a path on the surface is in the direction in which the gravitational potential drops fastest or equivalently the direction of the greatest negative gradient with distance of the height of the surface. Thus the gravitational forces that act upon the particle bear the same relationship with the gravitational potential as the electric force on a unit charge (the electric field) bears to the electrostatic potential.

Topics covered in Chapter 1

1. On an atomic or molecular scale the only significant forces are electromagnetic. At this most fundamental level all biological function, and indeed every other mechanism found in nature, must depend exclusively on such forces.

2. Two charges of like sign repel each other and two charges of opposite sign attract each other. The force exerted between the charges acts along the straight line joining the charges, is proportional to the product of the amplitudes of the two charges and is inversely proportional to the square of the distance between them. This is an example of 'action at a distance' and it operates even in a vacuum.

3. To deal with electrical forces between charges, the concept of an electric field is introduced. The electric field is represented by a vector. Associated with any positive charge in a vacuum is an electric field which points everywhere radially away from the

charge. Associated with every negative charge in a vacuum is an electric field which points radially towards the charge. The strength of the field in each case is proportional to the inverse square of the distance from the charge. When a second charge is introduced in the vicinity of the first charge, the second charge experiences a force, in the direction of the field created by the first charge, which has a magnitude equal to the product of the magnitude of the second charge and the strength of the field due to the first charge.

4.   Static electric fields originate and terminate only on charges. In a region where there are no charges, electric fields are continuous. Gauss's law states that, in a vacuum, the total outward flux of the electric field passing through a closed surface of arbitrary shape is equal to the total algebraic sum of the charges contained within that surface, divided by the constant number ε_0.

5.   The voltage difference between any two closely spaced points is defined as the energy required by an external agency to move a unit charge between these two points. Thus voltage times charge is a measure of energy and is a scalar quantity represented by a simple number. When the electric fields are created by static charges, the voltage difference between two points in space calculated along a given path does not depend upon the path chosen. As a result there is only one possible value for the voltage associated with any point in space, once the voltage has been defined at any other single location.

6.   The voltage at a given position is usually defined as the work done in moving a unit charge from a position very far away from any charges, to the given position. This definition depends on the assumption that the voltage everywhere, very far from all charges, is zero.

7.   The voltage and the electric field are connected in that the electric field at a given point and in a given direction is equal to the negative gradient of the voltage with distance at the same point and in the same direction.

Important equations of Chapter 1

1.   The electric field $E(R)$ at a distance R from a single charge $+Q$, located in a vacuum, acts in a radial direction away from the charge and has a magnitude given by the expression

$$E(R) = \frac{Q}{4\pi\varepsilon_0 R^2}.$$

2.   The force $\tilde{F}$ exerted on a charge q by an electric field $\tilde{E}$ is in the direction of $\tilde{E}$ and has an amplitude given by

$$F = qE.$$

3.   Gauss's law states that the total flux of the electric field out of any closed surface is equal to the algebraic sum of the charges contained within that surface, divided by the constant ε_0:

$$\iint_S E_N dS = \frac{\Sigma_i Q_i}{\varepsilon_0}.$$

For a continuous distribution of charge this equation may be written in terms of the charge density $\rho(x, y, z)$ in the elementary volume $dxdydz$ at the position (x, y, z) within the closed surface as

$$\iint_S E_N dS = \iiint_V \frac{\rho(x, y, z)}{\varepsilon_0} dxdydz.$$

4.   The voltage $V(x, y, z)$ is a scalar quantity defined as the work done by an external agency in moving a unit charge from infinity to the position defined by the coordinates (x, y, z). The voltage $V(x, y, z)$ is connected to the Cartesian components of the electric field (E_X, E_Y, E_Z) at the point (x, y, z) by the relations

$$E_X = -\frac{dV(x, y, z)}{dx}, \quad E_Y = -\frac{dV(x, y, z)}{dy}, \quad E_Z = -\frac{dV(x, y, z)}{dz},$$

or by the relation for the voltage rise between the points a and b,

$$V(b) - V(a) = -\int_a^b E_R dR.$$

The line integral of $\tilde{E}$, and hence the voltage difference $V(b) - V(a)$, is the same along any path connecting the points a and b.

5.   For static electric fields, the line integral of $\tilde{E}$ around any closed path is zero, which may be written in equation form as

$$\oint E_R dR = 0.$$

Problems

1.1   In a given region, the electric field points in the x-direction and has a constant magnitude of E_x. What can you deduce about the variation of the voltage $V(x, y, z)$ along the three Cartesian directions x, y and z? What additional information is required to establish the absolute magnitude of the voltage at any position?

1.2   A long straight wire carries a static charge of Q per unit length. At a radial distance R from the wire, and remote from its ends, what is the component of the electric field (a) in a radial direction and (b) parallel to the wire?

1.3   In the (x, y) plane two point charges, each with a charge Q, are placed at the positions $(0, a)$ and $(0, -a)$ on the y-axis. (a) By combining the electric fields of the two charges, deduce the electric field amplitude and direction at any point $(x, 0)$ on the x-axis. (b) Calculate the voltage at points $(x, 0)$ due to the two charges and hence deduce the direction and strength of the electric field anywhere on the x-axis from this voltage.

1.4   A sphere of copper with a radius of 1.193×10^{-2} m contains one gram mole or 6.022×10^{23} atoms of copper (the latter number being Avogadro's number). If the number of electrons differs from the number of positively charged copper atomic cores by as little as 1 part in 10^6, calculate the charge carried by the sphere. What is the force of repulsion between two such spheres placed a distance of 0.1 m apart? Assume that the force is the same as if the charge is all located at the centres of the spheres. Compare this force with the weight of 1 kilogram, which is 9.81 N.

1.5*   At a point P in space with Cartesian coordinates (x, y, z) the charge density is $\rho(x, y, z)$. By applying Gauss's law to a small cube at P of volume $dxdydz$ with its faces perpendicular to the x-, y- and z-axes, show that the components of the electric field at the point P must obey the relationship

$$\frac{dE_x}{dx} + \frac{dE_y}{dy} + \frac{dE_z}{dz} = \frac{\rho(x, y, z)}{\varepsilon_0}.$$

This is an important **microscopic statement of Gauss's law** that determines a relationship between the components of the electric field (E_x, E_y, E_z) at **a single point** and the charge density $\rho(x, y, z)$ at that point. This is in contrast to our macroscopic statement of Gauss's law in equation [1.7] which can only be applied over a finite volume. Note that this statement of Gauss's law can be written compactly using the divergence (see Appendix 1) of the electric field $\text{div}(\tilde{E})$ so that

$$\text{div}[\tilde{E}(x, y, z)] = \frac{dE_x}{dx} + \frac{dE_y}{dy} + \frac{dE_z}{dz} = \frac{\rho(x, y, z)}{\varepsilon_0}.$$

Worked solution

At the point $P(x, y, z)$ we construct an elemental cube of volume $dxdydz$ with its faces perpendicular to the three Cartesian axes. Let us consider the two faces of the cube that are perpendicular to the x-axis which each have an area $dydz$. Only the x-component E_x of the electric field $\tilde{E}$ field passes through these surfaces as the components E_y and E_z are tangential to the surfaces and do not penetrate them. Thus the total flux of the electric field $\tilde{E}$ out of the cube through these faces becomes the flux of the component E_x through these faces. The quantity $dE_x(x, y, z)/dx$ is the rate of variation of E_x

with x so that $[dE_x(x, y, z)/dx]dx$ gives the difference in the amplitude of E_x at the two opposite faces of the cube as in the equation

$$E_x(x + dx, y, z) - E_x(x, y, z) = \left[\frac{dE_x(x, y, z)}{dx}\right]dx.$$

Thus the total flux of the electric field component E_x out of the cube is given by

$$[E_x(x + dx, y, z) - E_x(x, y, z)]dydz = \left[\frac{dE_x(x, y, z)}{dx}\right]dxdydz.$$

Using similar arguments for the other four faces of the cube, we deduce that the **total flux** of electric field out of the cube is

$$\left[\frac{dE_x(x, y, z)}{dx} + \frac{dE_y(x, y, z)}{dy} + \frac{dE_z(x, y, z)}{dz}\right]dxdydz.$$

The total charge contained in the cube is $\rho(x, y, z)dxdydz$ so that, applying Gauss's law to the cube, we immediately obtain the required relation.

1.6   Using the expression derived in Problem 1.5, show that, in a region where there is no charge, an electric field which is all parallel to a given direction cannot vary in amplitude.

1.7   Using the expression derived in Problem 1.5 and the relationship connecting the electric field and the voltage expressed in equation [1.10], derive the **Poisson equation** which states that

$$\frac{d^2V(x, y, z)}{dx^2} + \frac{d^2V(x, y, z)}{dy^2} + \frac{d^2V(x, y, z)}{dz^2} = \frac{-\rho(x, y, z)}{\varepsilon_0}.$$

1.8   A spherical body of radius a bears a charge that is uniformly distributed throughout its volume with a charge density ρ per unit volume. Using Gauss's law, calculate the amplitude of the radial electric field $E_R(R)$ at a distance R from the centre of the sphere both inside and outside the sphere.

1.9*   Show that it is not possible, using only an electric field, to confine a charge in a fixed position in stable equilibrium in a region in which there are no other charges. This important fact is known as **Earnshaw's theorem**.

Worked solution

Let us choose a positive charge. If the charge is to be confined in stable equilibrium at a given position, the confining electric field must point inward towards that position from every direction. If we enclose the position in a Gauss law surface we would deduce that there must be negative charge within that surface because there is a flux of electric field entering the surface. But we are told that there are no other charges. Thus an electric field that converges to a given position from every direction is not possible in the absence of extra charges, and we have proved the theorem.

1.10   Prove Earnshaw's theorem using the result derived in Problem 1.7.

2. How conductors shape an electric field

Materials which contain charged particles that are free to move are called electrical conductors. Such material has a profound effect upon the shape and strength of electric fields both within the material and immediately around it. In this chapter the modification of a static electric field in a vacuum, which is caused by the presence of conductors, is discussed. Also discussed is an electric capacitor as an example of a charge storage device, and different geometries of capacitor are described. The plasma membrane of an animal cell is shown to act as a capacitor.

2.1 Electrical conductors

A **conductor** is defined as a material that contains charged particles that are able to **move freely** under the influence of an electric field. A metal is an example of a conductor because it contains electrons as mobile charged particles. The conductivity of biological fluids is usually due to the presence of charged mobile atoms or molecules in aqueous solution which are called **cations** if they are positively charged and **anions** if they are negatively charged. These ions move much more slowly than electrons and have very different characteristics which are important to understand when attempting to unravel biological function. As an example, the thermally activated diffusive motion of ions is such that the distance they diffuse grows only with the square root of the time elapsed rather that being proportional to that time. As a result, diffusion over very short distances may be rapid but over long distances it becomes prohibitively slow, necessitating special mechanisms, such as exist within nerve cells, for efficient electrical signalling over large distances. We will discuss some characteristics of diffusive ionic motion in Chapter 3 and some other ionic properties in Chapters 11, 12 and 13, but here, for a discussion of the shaping of electric fields by conductors, we need only know that conductors contain free mobile charge carriers.

As mentioned above, the model of a conductor most relevant to biology is an aqueous solution that contains cations and anions which are free to migrate through the solution. If we were able to penetrate into the interior of the solution and pass close to an ion we would experience a very large electric field created by the charge on the ion. Such internal fields we call **local electric fields**. If we take as a model a solution of a univalent–univalent salt such as KCl at a concentration of 0.1 molar (where 1 molar denotes Avogadro's number of particles – recall Problem 1.4 – in a litre of solution), the average distance between a cation and a neighbouring anion is only about 10^{-9} m. Thus the local electric field will vary in sign and direction over these microscopic distances. To demonstrate that these local fields average to zero we can consider a neutral solution located in a region where there are no external electric fields applied. We then calculate the line integral of the electric field $\oint E_R dR$ around a closed path which passes through the solution. We know from Equation [1.12] that when taken around any complete circuit, $\oint E_R dR = 0$, and we have also arranged that the electric field is zero everywhere outside the solution. Thus we can deduce that $\int E_R dR = 0$ over the part of the integration path that passes through the solution. The electric field used to calculate the integral will of course include the local electric fields so that we are able to deduce that, over **any path** through the solution, the component of $\tilde{E}$ in the direction of the path E_R sums to zero. We can then say that the **average electric field** within the neutral conducting solution is zero. We are distinguishing here between the internal **local field** that varies markedly over microscopic distances and the **average field** which is an average taken over a macroscopic distance. Although the electrostatic laws we have deduced in Chapter 1 hold for all electric fields, including local fields, we are most often interested in the average field, which is an average of the electric field over a macroscopic volume. In this chapter we will refer to the average electric field inside a conductor, but in later chapters we will simply refer to the electric field within a conductor, knowing that it is an average over a macroscopic volume that is intended.

2.2   The electric field within a conductor

We will take as an example of a conductor, a sphere of aqueous solution containing some mobile cations of charge $+Q_c$ and anions of charge $-Q_a$ and surrounded by a non-conducting barrier. This may be taken as a very crude model of an animal cell with the containing non-conducting barrier representing the cell plasma membrane, as sketched in Fig. 2.1a.

If an electric field $\tilde{E}$ is applied to the sphere, cations with charge $+Q_c$ will experience a force $+Q_c E$ in the direction of the field and anions of charge $-Q_a$ will experience a force $-Q_a E$ in the direction of the field or, equivalently, a force of $+Q_a E$ in a direction opposite to the field, as discussed in Section 1.3. These charges will move until arrested by the cell membrane. The end result will be a layer of positive charge (the arrested cations) on the inner face of the membrane at one side of the cell

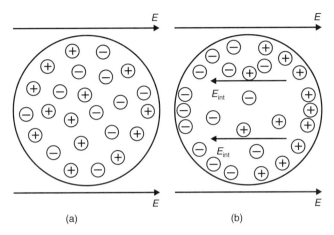

Fig. 2.1   (a) A spherical body containing mobile positive and negative charged particles is placed in a static electric field $\tilde{E}$. (b) Under the influence of the electric field, some of the charged particles move until arrested by the boundaries of the body. These accumulating charges create an internal electric field $\tilde{E}_{int}$ which exactly cancels to zero the applied field $\tilde{E}$, so that the total electric field is zero everywhere within the conducting body.

and a layer of negative charge (the arrested anions) on the opposite inner membrane face. These layers of surface charge will create a secondary internal electric field $\tilde{E}_{int}$ which is directed such as to oppose the original applied field $\tilde{E}$, as sketched in Fig. 2.1b. Clearly this process will continue until the average electric field everywhere within the body of the cell is exactly zero, at which time mobile ions will no longer experience an electrical force and will cease their motion. If a residual average field exists anywhere within the conductor, further charges will move under its influence until the internal average field is cancelled exactly to zero. This is clearly true for a conductor of any shape which is surrounded by an insulating boundary. Thus we arrive at the important result that **if a static external electric field is applied to a conductor, charges will migrate within the conductor such as to cancel to zero the average electric field within the body of the conductor**. Note that for this argument to hold the conductor must contain charges that are **free to move**. Were the mobile charges to be tethered within the cell in any way then the argument would not hold. The charges would move until the combined force of the electric field and the restraining force of the tether summed to zero, but this would not ensure zero average field within the conductor. Another important consequence of the absence of average electric fields within conductors is that the voltage everywhere within a conductor must be the same. This can be seen by the application of equation [1.11] which asserts that

$$V(b) - V(a) = -\int_a^b E_R dR.$$

Conductors are thus equipotential bodies in that the electrostatic potential or voltage is constant throughout the volume of the conductor and on its surface.

This cancellation to zero inside a conductor of an externally applied static electric field is called **screening** and has important implications for animals exposed to such fields. A whole animal may be considered on a macroscopic scale as an insulating membrane containing conducting fluid which leads to the first attenuation of any static electric field applied outside it. Within the animal every closed compartment such as a cell gives rise to further screening of any electric field that may be created within the animal but outside the cell. Thus within a cell, inside an animal, the only remaining average static electric field we would expect to find is that which is generated within that cell, and any field created by the externally applied static electric field will be negligibly small.

Note that the screening applies to the bulk of the interior of the conductor but not very near its insulating boundaries where the mobile charge carriers accumulate. Very near these charged surfaces there is an average electric field which points in a direction normal to the surface. This gives rise to a force which is always outward, no matter what the sign of the accumulated charges, and is due to like charge

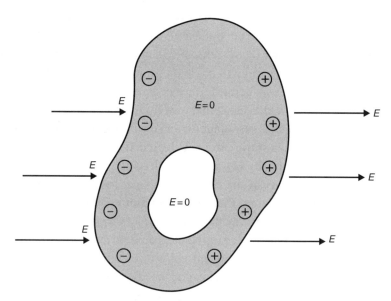

Fig. 2.2   A conductor of arbitrary shape in an electric field $\tilde{E}$. The motions of the mobile charges within the conductor create internal electric fields which ensure that the total electric field within the conductor is everywhere cancelled to zero and the interior of the conductor is an equipotential region. Any cavity carved within the conductor is also an equipotential region so that the electric field within the cavity is necessarily zero.

repulsion. This force is balanced by that exerted by the containing insulating boundary layer. A second limitation to the effectiveness of electrical screening results from the finite speeds with which the mobile ions move under the influence of the applied fields. If an external electric field is applied abruptly it does initially penetrate the conductor without much attenuation but it then decays with time in response to the motion of the internal charged ions as described above. For typical animal soft tissue the initial field within a cell decays by a factor of 2.7 in a time of one thousand-millionth of a second (10^{-9}s). This limits the effectiveness of the screening of electrical fields which are rapidly switched off and on or which alternate rapidly with time. In this section the discussion has been limited to the screening of time-independent or static electric fields.

Screening by a conductor is also a useful practical experimental technique to eliminate static electric fields. We consider a conductor of any shape that contains within it a cavity of any shape, as sketched in Fig. 2.2. Because the interior of the conductor is an equipotential, the potential everywhere on the walls of the cavity is the same. Thus, if there is no charge located within the cavity, the whole cavity is at a constant potential so that the electric field within the cavity, which is the negative gradient of the potential, must be zero. Thus surrounding an item of experimental equipment with a seamless metal box ensures that no external static electric field penetrates the box to act upon the equipment.

2.3   The electric field surrounding a conductor

A second effect of conductors on the shape of an electric field may be deduced by considering the interface between a conductor and a surrounding vacuum, as sketched in Fig. 2.3a. We will label the component of the electric field parallel to the surface of the conductor and just outside it as E_{out}^P and the component parallel to E_{out}^P but just on the inside as E_{in}^P. We construct a rectangular path $abcda$ as in Fig. 2.3a, with two sections of the path ad and bc parallel to the surface with length l and two very short sections ab and cd of length m perpendicular to the surface. We may then use the fact that the line integral of the electric field around any closed circuit is zero as in equation [1.12]. The contribution to the line integral around the circuit $abcda$ from the two sections ab and cd, from electric fields perpendicular to the surface, may be made smaller by reducing their length m. However, because there is charge in the surface of a conductor there can be large fields perpendicular to the surface as it is crossed which can make an appreciable contribution to the integral even for a short path length. Fortunately, when we calculate the line integral around the complete path $abcda$ any surface contributions from the path sections ab and cd cancel as they are traversed in opposite directions relative to the surface. Because the field amplitude will not

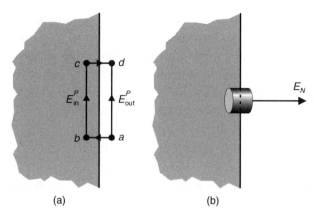

Fig. 2.3   (a) The component of the electric field parallel to the surface of a conductor and just outside it is shown as E_{out}^P. The parallel component just inside the conductor is shown as E_{in}^P. By using the fact that the line integral of the electric field around the complete circuit $abcda$ is zero, we may deduce that $E_{\text{out}}^P = E_{\text{in}}^P$. From the fact that E_{in}^P, a field within a conductor, is zero we deduce that E_{out}^P is zero, so that all electric fields must meet conductors normal to the conducting surface. (b) A small cylindrical surface which penetrates the surface of a conductor with its axis normal to the surface of a conductor. By applying Gauss's law to this cylinder, a relationship is obtained between the surface charge density of the conductor and the component of the electric field E_N outside the conductor and normal to its surface.

vary appreciably within the small distance l, the line integral reduces to the simple expression

$$0 = \oint_{abcda} E_R dR = \left(E_{\text{in}}^P - E_{\text{out}}^P\right)l \quad \text{or} \quad E_{\text{in}}^P = E_{\text{out}}^P. \qquad [2.1]$$

But we know that $E_{\text{in}}^P = 0$, because all average electric fields within a conductor are zero, so that we may deduce that necessarily $E_{\text{out}}^P = 0$. By this argument we have shown that the component of any electric field external to a conductor and parallel to its surface is zero, or equivalently, that **all static electric fields must meet the surface of a conductor at right angles**.

More information about the electric fields around a conductor may be deduced by constructing a very short right-circular cylinder with its flat faces parallel to the surface of the conductor such that the cylinder encloses part of the surface of the conductor, as in Fig. 2.3b. If the area of the flat ends of the cylinder is A and the density of charge per unit area of the surface is Q_s, then the total charge included in the cylinder is $A \cdot Q_s$. No electric field penetrates the flat surface of the cylinder within the conductor as the field there is zero and no field cuts the curved walls of the cylinder because, as shown above, the field is perpendicular to the conductor surface and thus parallel to the cylindrical wall. If we apply Gauss's law to the cylinder, the only flux of electric field is that through the outer flat surface and we obtain the equation

$$AE_N = A\left(\frac{Q_s}{\varepsilon_0}\right) \quad \text{or} \quad E_N = \frac{Q_s}{\varepsilon_0}. \qquad [2.2]$$

where E_N is the outward-pointing electric field which must be normal to the surface, thus relating the electric field near a conductor with the charge density on its surface. It is this field E_N acting on the surface charge density Q_s that gives rise to the out-ward force at the surface mentioned above.

Equation [2.2] is correct but it can give the impression that the only contribution to E_N is from the very local surface charge density within the Gauss law cylinder in Fig.2.3b. Why does the charge on the surface of the conductor outside the cylinder or any other applied fields not contribute to E_N? It is worth deriving equation [2.2] another way to make this clear. Consider the layer of surface charge of surface density Q_s and neglect the fact that it is on the surface of a conductor. If we constructed a Gauss law cylinder as in Fig. 2.4 enclosing the surface, we would deduce that the total flux of the electric field E_s out of each of the flat ends of the cylinder is AE_s. Here E_s is

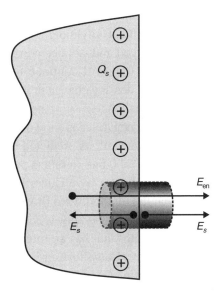

Fig. 2.4   A small cylindrical Gauss law surface which penetrates the surface of a conductor and has its axis normal to the surface. The surface charge of the conductor is considered as a plane sheet of charge with a surface density Q_s producing normal and local electric fields E_s in both directions away from the surface. All other electric fields at the cylinder, from applied electric fields and remote surface charges, produce a resultant environmental electric field of E_{en}. By invoking the fact that E_s and E_{en} must sum to zero inside the conductor, the total normal electric field E_N just outside the conducting surface is deduced.

the electric field which is perpendicular to the surface charge layer and points away from the layer in **both directions**, outward like E_N but also inward. We can invoke the symmetry of the locally flat charged layer to deduce that the two oppositely directed components E_s have the same amplitude and are perpendicular to the surface, as shown in Fig. 2.4. Applying Gauss's law to the cylinder, we then deduce that the amplitude of the fields E_s created by the surface charge, is given by the equation

$$2AE_s = A\left(\frac{Q_s}{\varepsilon_0}\right) \quad \text{or} \quad E_s = \frac{Q_s}{2\varepsilon_0}. \tag{2.3}$$

We must add to E_s the electric field created by the charges located on the surface of the conductor but outside the Gauss law cylinder and also any other external applied fields. We call this field, due to the electrical environment, E_{en}. We can then invoke the fact that we know that the total average field in the conductor is zero to deduce that the total field penetrating the inner flat face of the Gauss law cylinder is zero. This means that within the conductor and at its surface, $\tilde{E}_{en}$ is directed oppositely to $\tilde{E}_s$ and has the same amplitude so that $E_{en} = E_s = Q_s/2\varepsilon_0$, and the two fields cancel. At the outer flat surface of the Gauss law cylinder $\tilde{E}_{en}$ is parallel to $\tilde{E}_s$ so that $E_N = E_{en} + E_s = Q_s/\varepsilon_0$ as in equation [2.2]. In this manner it is seen that, despite the apparent simplicity of the Gauss law derivation of equation [2.2], there are contributions to the field E_N from the electrical environment outside the Gauss cylinder.

Note that if we consider an uncharged and isolated conductor placed in an applied electric field such as that in Fig. 2.1, it will continue to carry no total charge in the electric field as its isolation precludes any change in its total charge. However, the electric field will rearrange its internal charge such that near the surface some parts become positively charged and some negative, as shown in Fig. 2.1b. As a result, the electric field external to the conductor will, as predicted by equation [2.2], in some areas point into the surface where $Q_s < 0$ and in others point away from the surface where $Q_s > 0$. If we enclose the conductor by a Gauss law surface that closely fits the conductor we may deduce that the total outward flux of electric field through this surface is zero as the total charge on the conductor is zero.

Finally, we will consider a conductor that has an **uncompensated** total charge of Q_{tot}. If we enclose this conductor with a Gauss law surface that closely fits its surface, but is just outside it, we may deduce using Gauss's law that the total outward flux of the electric field through this surface is equal to the total uncompensated charge in and on the conductor, divided by ε_0. However, this total flux of electric field is the same as the sum of all fluxes due to surface charges like that used when deducing Equation [2.2]. We may then deduce that for a charged conductor the total uncompensated charge Q_{tot} **all resides on the surface of the conductor**. This makes good sense physically as an assembly of uncompensated charges of the same sign repel each other and, for the lowest energy equilibrium state, they seek to maximize the distances between them. This ensures that they all reside in the surface of the conductor.

The properties of conductors are remarkable in that, no matter how much uncompensated charge is placed on the conductor and no matter what external electric fields it is exposed to, all the uncompensated charge must reside in the surface layer, and the total surface charge distributes itself such that there are no fields within the conductor and all external fields meet the surface of the conductor at right angles.

2.4   The electrical capacitor

Imagine two plane and parallel circular conducting plates of large area A separated from each other by a fixed small distance d and sharing a common axis, through their centres and normal to their planes, as in Fig. 2.5. We suppose that we place an uncompensated charge of $+Q_{tot}$ on the upper plate and an uncompensated charge of $-Q_{tot}$ on the lower plate. The attraction between the charges of opposite sign ensures that nearly all the charge resides on the inner surfaces of the two plates. We know from our discussion above that the electric field near the two plates is perpendicular to the plates and the symmetry of the system demands that the electric field between the plates is also everywhere perpendicular to the plates except possibly near their edges. If we consider the cylindrical Gauss law surface marked 1 in Fig. 2.5 which has its

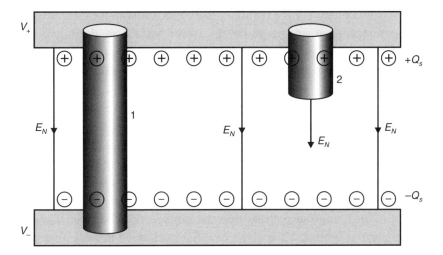

Fig. 2.5   A parallel plate capacitor with surface charge densities $+Q_s$ and $-Q_s$ on the inside surfaces of its upper and lower plane conducting plates. The voltage difference between the upper and lower plate is $V_+ - V_-$ and the internal electric field, normal to the plates, is E_N. Gauss's law is applied to two cylindrical surfaces marked 1 and 2 in the diagram, which have their axes normal to the plates, to deduce the properties of the capacitor.

plane ends buried in the two conductors and its axis perpendicular to the plates, we can deduce that the flux of the electric field out of the cylindrical surface is zero. None penetrates its flat ends because they are buried within conductors where the average electric field is necessarily zero, and none penetrates its curved surface because the field is parallel to that surface. Thus we deduce from Gauss's law that the total charge within the cylinder is zero or that the excess charge density (charge per unit area) on the lower surface of the upper plate $Q_s = Q_{tot}/A$ is exactly equal in magnitude and opposite in sign to that on the upper surface of the lower plate.

We now apply Gauss's law to the cylindrical surface marked 2 in Fig. 2.5 with one flat end of area S buried in the top conductor, the other between the plates, and with the cylinder axis perpendicular to the plates. The uncompensated charge within the cylinder all lies on the lower surface of the upper plate and is equal to SQ_s so that, as in equation [2.2], the electric field leaving its lower flat surface is given by

$$E_N = \frac{Q_s}{\varepsilon_0}. \tag{2.4}$$

Essentially because there are no charges in the gap between the plates, we would obtain the same value of field in equation [2.4] everywhere in the gap. We can deduce this by considering Gauss law cylinders of different lengths, terminating at different places within the gap, which all predict the same field E_N through the lower plane face of the cylinder. This means that the field is constant in magnitude between the plates. Thus we arrive at a picture of the parallel plate capacitor with the field originating in the uncompensated positive charge on the lower surface of the upper plate and terminating on the uncompensated negative charge on the upper surface of the lower plate. In the region between the plates the electric field is always perpendicular to the plates.

Because the field is constant in magnitude, the voltage difference between the upper and lower plates is easily calculated using equation [1.11]. Particular attention must be paid to the sign of the voltage change. Equation [1.11] gives the voltage change as the path is traversed, so that the line integral $- \int E_R dR$ along the given path gives $V(\text{end}) - V(\text{start})$. The voltage always drops in the positive direction of the electric field. If we wish instead to calculate $V(\text{start}) - V(\text{end})$ we need to form the line integral $+ \int E_R dR$ along the path so that the voltage drop between the positively charged top plate at voltage V_+ and the negatively charged bottom plate at voltage V_- is given by $V_+ - V_-$ as follows:

$$V_+ - V_- = + \int_{top}^{bottom} E_N dR = E_N \int_{top}^{bottom} dR = E_N d = \frac{Q_s d}{\varepsilon_0}. \tag{2.5}$$

Writing equation [2.5] in terms of the total uncompensated charge on the top plate Q_{tot}, rather than the surface charge density Q_s, which is a charge per unit area, we obtain

$$V_+ - V_- = \frac{Q_{tot}d}{A\varepsilon_0} \quad \text{or} \quad \frac{Q_{tot}}{V_+ - V_-} = C = \frac{A\varepsilon_0}{d}. \qquad [2.6]$$

The parameter C measures the charge stored on each plate per unit voltage difference between the plates and is called the **capacitance** of the capacitor. It is expressed in units called farads, written simply as F, and it measures the effectiveness of the capacitor as a charge storage device. In the case of the parallel plate capacitor it is directly proportional to the area of the plates and inversely proportional to their separation.

When deducing the properties of this type of capacitor we have idealized the situation in that we have ignored effects of field curvature near the edges of the plates and the fact that some field lines will originate from uncompensated charge on the outer edges of the plates and even the outer surfaces of the plates near these edges. However, if the plates have a large area and are close together, these corrections are small.

2.5   The animal cell plasma membrane as a capacitor

The outer plasma membrane of an animal cell may be considered as a capacitor in that it is a parallel-sided non-conductor layer which separates two conductors, namely the surrounding conducting aqueous fluid and the conducting fluid contained within the cell. In fact the capacitance of a typical animal cell outer plasma membrane is about $10^{-6}\,\mathrm{F\,cm^{-2}}$ or, in SI units, $10^{-2}\,\mathrm{F\,m^{-2}}$. This is in fact a very large capacitance per unit area because the thickness of the membrane d is so small. We will now make a rough estimate of the voltage difference that results from the transfer of some ions across the membrane. We take as a simple model of a cell a sphere of radius 5 micrometres containing a solution of 0.1 molar potassium chloride and surrounded by a typical animal plasma membrane. The surface area of the cell is about $3.14 \times 10^{-10}\,\mathrm{m^2}$ so that the total capacitance C of the membrane is about $3.14 \times 10^{-12}\,\mathrm{F}$. To generate an easily measured voltage of say $dV = 5$ millivolts across the membrane we would have to transfer a charge of $Q = CdV = 1.57 \times 10^{-14}\,\mathrm{C}$ across the membrane which corresponds to 9.82×10^4 univalent cations each with the elementary charge of $1.6 \times 10^{-19}\,\mathrm{C}$. But a cell with a radius of 5 micrometres and containing 0.1 molar KCl contains 3.15×10^{10} univalent potassium cations. Thus to generate an easily measured voltage across the membrane of 5 millivolts requires a change in the number of cations within the model cell of only 3 parts per million, a concentration difference far too small to be detected in any other way. This is an illustration of the strength of electric effects for, despite the fact that the membrane has a very large capacitance and thus a very large capacity to store charge with little voltage rise, the electrical effect of the transfer of the ions across the membrane is much easier to measure than any other macroscopic consequence of the transfer.

2.6   The capacitance of a single conductor

I have described a capacitor consisting of two closely spaced conductors. Here we will consider the capacitance of a single conductor with the second conductor situated a very long way away. We take as an example the simplest geometry of a single isolated spherical conductor of radius a carrying a total uncompensated charge of $+Q$, as shown in Fig. 2.6. If we construct a spherical Gauss law surface of radius R with $R > a$ centred on the centre of the sphere, we deduce by symmetry that the electric field at a fixed value of R, which we represent as $E(R)$, around this sphere is everywhere constant in amplitude and directed radially out of the sphere. Applying Gauss's law, we find that the total flux of the electric field out of the Gauss surface is given

$$J_{tot}(R) = 4\pi R^2 E(R) = \frac{Q}{\varepsilon_0} \quad \text{or} \quad E(R) = \frac{Q}{4\pi\varepsilon_0 R^2}. \qquad [2.7]$$

The first deduction is that the electric field $E(R)$ created by a spherical conductor carrying a uncompensated charge of $+Q$ is everywhere radially pointing outward and, comparing equation [2.7] with equation [1.2], it is the same as if all the charge were concentrated at the centre of the sphere. All the electric field that originates on the surface of the sphere must end on negative charges situated infinitely far away where $V(\infty) = 0$. The change in voltage between the equipotential conducting sphere and the zero of voltage at infinity may be deduced using equation [1.11] with a sign change as we discussed in deriving equation [2.5]. Thus we have

$$V(a) = V(a) - V(\infty) = + \int_a^\infty E(R)dR = +\left(\frac{Q}{4\pi\varepsilon_0}\right) \int_a^\infty \frac{dR}{R^2} = +\frac{Q}{4\pi\varepsilon_0 a}. \qquad [2.8]$$

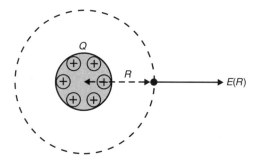

Fig. 2.6   A single spherical conductor with radius a and carrying a total charge Q. The electric field $E(R)$ at a distance R from the centre of the spherical conductor is shown.

Thus the capacitance of a conducting sphere of radius a is given by

$$C = \frac{Q}{V(a)} = 4\pi\varepsilon_0 a. \qquad [2.9]$$

We have already defined the zero of voltage as that at infinity, far from any charges so that $V(\infty) = 0$. However, in electrical problems a second definition of zero voltage is often used. We define the voltage of the Earth, considered as a conducting sphere, as zero and thus the voltage of any conductor electrically connected to the Earth to be zero. The capacitance of a conducting sphere the size of the Earth is in fact surprisingly small, as shown in Problem 2.7. The fact that the voltage of the Earth is relatively stable has more to do with the cancellation of the positive and negative charges applied to it than the fact that its capacitance is very high.

2.7   The spherical capacitor

Finally, we will discuss another type of capacitor that consists of a central conducting sphere of radius a surrounded by a second concentric conducting spherical shell of inner radius b $(b > a)$ which is connected to the Earth as sketched in Fig. 2.7. We

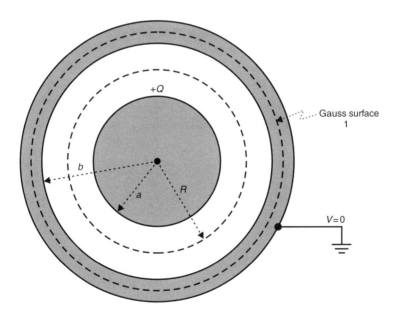

Fig. 2.7   A spherical capacitor consisting of a spherical conductor of radius a which carries a total charge $+Q$ and a concentric conducting shell with an inner radius b which is at zero voltage because it is connected to Earth. Two spherical Gauss law surfaces are shown, one within the thickness of the outer spherical shell and another of radius R within the cavity between the conductors.

place an uncompensated charge of $+Q$ on the inner sphere. If we construct a spherical Gauss surface like that marked as 1 in the figure which is concentric with the capacitor but located within the outer conducting shell, we may deduce that the flux of the electric field out of this surface is zero, simply because there exists no average electric field within a conductor. Thus we deduce that the charge on the inner surface of the outer shell is exactly equal to $-Q$ so that this Gauss surface contains no uncompensated charge. Physically this negative charge on the outer shell has been attracted to the positive charge of $+Q$ on the inner sphere and flows from the Earth along the wire connecting it to the Earth, to reside on the inner surface of the outer shell. We may deduce that there is no uncompensated charge on the outer surface of the shell by constructing a Gauss law surface as a sphere completely enclosing the capacitor. There is no electric field between the outer shell at zero voltage and infinity at zero voltage because everywhere outside the sphere is at the same voltage. Thus the flux of the electric field cutting this outer Gauss law surface is zero so that, using Gauss's law, we may deduce that there is no uncompensated charge inside this outer Gauss law surface. The only uncompensated charges are $+Q$ on the outer surface of the inner sphere and the charge of $-Q$ on the inner surface of the outer shell. As with the parallel plate capacitor where the parallel electric field that originated in the uncompensated positive charges on the upper plate and terminated in the equal but negative charges on the lower plate, here the radial field that originates in the positive uncompensated charges on the outer surface of the inner sphere terminates on the equal-amplitude negative charge on the inner surface of the outer shell.

If we construct a spherical Gauss law surface of radius R between the two conductors we may deduce that the electric field at radius R is given by $E(R) = Q/4\pi\varepsilon_0 R^2$ as in equation [2.7]. Integrating along a radial path between the two conductors from $R = a$ to $R = b$, we find that

$$V(a) - V(b) = + \int_a^b E(R)dR = \frac{Q}{4\pi\varepsilon_0} \int_a^b \frac{dR}{R^2} = \frac{Q}{4\pi\varepsilon_0}\left(\frac{1}{a} - \frac{1}{b}\right),$$

so that the value of the capacitance created by the two charged spherical conductors is given by

$$C = \frac{Q}{V(a) - V(b)} = 4\pi\varepsilon_0\left(\frac{ab}{b-a}\right). \qquad [2.10]$$

Note that this result becomes equal to that in equation [2.9] when b becomes very large in comparison with a.

Topics covered in Chapter 2

1.   A conductor is defined as a body which contains charged particles that are free to move under the influence of applied electric fields.

2.   When a conductor is placed in a static electric field some of the mobile charges will move under the influence of the applied field so as to reduce the average total electric field within the body of the conductor to zero. However, there does exist an electric field pointing normally to the surface very close to that surface.

3.   The fluid interior of an animal cell is a conductor which is insulated from the conducting fluid outside by the non-conducting plasma membrane. If a static electric field is applied externally to such a cell the average electric field within the body of the cell is reduced to zero by the motion of the mobile ions within the cell. This effect is called electrical screening; it greatly reduces the effects within a cell of externally applied electric fields.

4.   In equilibrium, when the charge motion within a conductor has terminated, the electric field just outside a conductor placed in an externally generated static electric field is everywhere directed perpendicularly to the surface of the conductor.

5.   The charge storage capacity of an isolated conductor is measured by its capacitance, which is defined to be the ratio of the charge it carries to its voltage.

6.   Purpose-built charge storage devices or capacitors often consist of two conductors placed close together and charged to different voltages. One example is the parallel plate capacitor which consists of two identical plane horizontal conducting plates placed one above the other with a constant small separation between the plates. If the lower plate is connected to a large body such as the Earth and the upper plate is charged to a higher voltage and then isolated, it is found that the charges residing on the inside surfaces of the two conductors are equal in magnitude but opposite in sign. The capacitance of such a device is given by the ratio of the magnitude of the total uncompensated charge stored on either plate to the voltage difference between the two plates.

7.   The thin non-conducting plasma membrane of an animal cell, sandwiched between the conducting fluid that bathes its inner and outer surfaces, may be considered as a capacitor. When charged ions move across the membrane the most easily detected macroscopic effect is a change in the trans-membrane voltage.

Important equations of Chapter 2

1.   In electrostatics the average electric field within a conductor E_{int} is zero. As a result the voltage within the conductor V_{int} is constant:

$$E_{int} = 0 \quad \text{and} \quad V_{int} = \text{constant}.$$

2.   Just outside the surface of a conductor placed in a static electric field all components of the electric field parallel to the surface E_p are zero:

$$E_p = 0.$$

As a result of this, all external static electric fields must meet conductors perpendicular to the conductor surface.

3.   All the uncompensated charge within a conductor resides very near the surface.

4.   The capacitance C of a conductor measures the ratio of the total uncompensated
charge on the conductor Q and its voltage V so that

$$Q = CV.$$

5.   The capacitance of a capacitor formed from two identical plane parallel conducting plates of large area A, placed one above the other and separated by a small distance
d, is given by

$$C = \frac{A\varepsilon_0}{d}.$$

6.   The capacitance of a capacitor formed by a conducting sphere of outer radius a
and a concentric conducting spherical shell of inner radius b ($b > a$), in which the
inner sphere is charged and the outer shell is connected to Earth, is given by

$$C = 4\pi\varepsilon_0 \left(\frac{ab}{b-a} \right).$$

7.   The capacitance of a single charged sphere of radius a suspended in isolation in
a vacuum is given by

$$C = 4\pi\varepsilon_0 a.$$

Problems

2.1   A parallel plate capacitor is formed from three plane circular conducting plates
each of area A. The plates are arranged one above the other with their planes parallel
and have a common axis passing through the centres of the discs. The distance
between neighbouring discs is d. The top and bottom discs are connected to Earth
and a charge $+Q$ is placed on the middle disc. What is the voltage on the middle disc
and the capacitance formed between the middle and the two outer discs?

2.2   A conducting sphere of radius a is first connected to Earth, it then touches
a conductor maintained at a voltage V and finally moves away from it. How much
charge has been removed from the conductor at the voltage V?

2.3   A thin spherical conducting shell of radius a carries a total charge Q. Calculate the dependence upon R, the distance from the centre of the shell, of (a) the electric field and (b) the voltage both inside ($R < a$) and outside ($R > a$) the shell. Sketch the dependence of the field and the voltage as a function of R. Why must the voltage curve always be continuous with no abrupt changes?

2.4   The spherical capacitor discussed in Section 2.7 had a charge Q on the inner sphere while the outer spherical shell was connected to the Earth. If, instead, the charge Q is placed on the outer shell and the inner sphere is connected to the Earth, the capacitance is changed. Explain as fully as you can the reason for the change.

2.5   The plasma membrane of an animal cell bears a voltage of $70\,\mathrm{mV}$ across it. Assume that this voltage difference is created by surface charge densities of $+\sigma_s$ and $-\sigma_s$ on its opposite faces. The capacitance of such a membrane is typically $10^{-2}\,\mathrm{F\,m^{-2}}(1\mu\mathrm{F\,cm^{-2}})$. Find σ_s and show that it is equivalent to a single univalent charge $(1.6 \times 10^{-19}\,\mathrm{C})$ for each square area of the surface with a side of about $15 \times 10^{-9}\,\mathrm{m}$.

2.6   The outer plasma membrane of the spherical cell of radius $5\,\mu\mathrm{m}$ discussed in Section 2.5 has a capacitance of $C = 3.14 \times 10^{-12}\,\mathrm{F}$, a thickness of $5 \times 10^{-9}\,\mathrm{m}$, and has a voltage difference of $V = 50\,\mathrm{mV}$ across it. What is the amplitude of the electric field that acts across the membrane? Calculate the electrostatic energy $\frac{1}{2}CV^2$ stored in the charged membrane and compare this with (1) the change in energy when a single univalent ion ($q = 1.6 \times 10^{-19}\,\mathrm{C}$) crosses the membrane and (2) a typical thermal energy at $20\,^{\circ}\mathrm{C}$ of $kT = 4.046 \times 10^{-21}\,\mathrm{J}$.

2.7   The Earth may be considered as a conducting sphere of radius $6370\,\mathrm{km}$ suspended in a vacuum. What is its capacitance in farads?

3. Ionic conductors

When a voltage is applied across the ends of a conductor an electrical current is caused to flow. In this chapter Ohm's law, which governs the current flow, is discussed and the concepts of conductivity and resistivity of a conducting material are introduced. In biology the most important conductors consist of liquids which contain ions which are mobile particles bearing charges. The laws which determine the thermally activated passive diffusion of such particles are discussed, as is the driven motion when an electric field is applied.

3.1 Ohm's law of conductivity

In this chapter we will deal with some of the properties of conductors as current carriers and describe some of the unusual properties of ions in solution which are the dominant current carriers in biology. It is found experimentally for a conductor of fixed dimensions at a constant temperature that the electric current that flows through it is directly proportional the voltage applied between its ends. Electric current is defined as the positive charge that passes through a given cross-section of the conductor each second and is measured in units of coulombs per second ($C\,s^{-1}$). When describing a current, the flow of $+Q$ coulombs per second of positive charge in a given direction is exactly equivalent to the flow $-Q$ coulombs per second of negative charge in the opposite direction. The experimental finding is called **Ohm's law** and the constant of proportionality connecting the voltage applied between the ends of the conductor $V_1 - V_2$ and the current flowing I is called the **resistance** of the conductor R, so that

$$V_1 - V_2 = RI. \qquad [3.1]$$

The voltage drop $V_1 - V_2$ along the conductor provides the energy for the current flow, just as the pressure drop between the ends of a pipe provides the energy for liquid flow along the pipe. Note that along a conductor carrying a current there are voltage differences implying electric fields within the conductor. This would seem to be a violation of the previous statement that conductors are all at the same voltage, and that the average electric field within a conductor is zero. The difference is that in the case of current flow the conductor is not in equilibrium. A conductor becomes an equipotential body in equilibrium only when all charge migration has terminated.

The characteristic **resistivity** of a given substance is defined as the resistance between opposite faces of a metre cube of that substance and is often represented by the symbol ρ. The resistance R of a length L of this substance with a constant cross-sectional area of A is then given by

$$R = \rho\left(\frac{L}{A}\right). \tag{3.2}$$

If the mobile charge carriers could move freely we would not expect a relation like Ohm's law to hold. The voltage difference implies an electric field within the conductor which, by Newton's second law, would accelerate the charge carriers so that the current would increase rapidly with time rather than remaining constant. To obey Ohm's law the charge carriers must move with a constant average velocity in a constant negative voltage gradient or, equivalently, in a constant electric field. The explanation lies in the fact that the charge carriers are continuously being accelerated by the electric field but almost immediately colliding with other molecules or atoms and returning to rest. The energy given to the charge carrier by the electric field is lost in the collisions as heat. In a metal the charge carriers are electrons, and they collide with the atoms of the static metallic crystal lattice converting their kinetic energy into heat. The velocity in the x-direction, v_x, of a positively charged particle in a constant electric field E_X in the same direction is illustrated in Fig. 3.1. The particle is accelerated at a constant rate for a variable time t until it collides with another atom or molecule and its velocity v_x is reduced to zero. The durations of the times t are random, which results in a constant average drift velocity v_d in the x-direction.

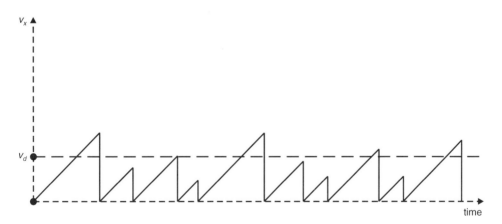

Fig. 3.1 A graph plotted against time of the x-component of the velocity of a mobile charged particle v_x which is continuously accelerated by a constant electric field for a variable time t before v_x is brought to zero by collision with a static body. The time-averaged drift velocity v_d in the x-direction which results from this motion is shown.

Charged ions in aqueous solution, as we shall discuss in some detail, continuously collide with neighbouring water molecules and dissipate to the solution as heat the energy gained by acceleration in the electric field.

Another parameter useful when describing a conductor is its conductivity σ, which is defined as the reciprocal of its resistivity, $\sigma = 1/\rho$. Let us consider an infinitesimal cube of material defined in Cartesian coordinates to have a volume $dxdydz$ with a voltage drop $-dV_x$ between its two faces perpendicular to the x-axis. Applying equations [3.1] and [3.2] to this cube, we obtain

$$-dV_x = RI_x \quad \text{or} \quad -dV_x = \left(\frac{\rho dx}{dydz}\right)I_x = \frac{1}{\sigma}\left(\frac{dx}{dydz}\right)I_x.$$

Rearranging the parameters, we obtain

$$\frac{I_x}{dydz} = -\sigma\frac{dV_x}{dx} \quad \text{or} \quad J_x = \sigma E_x. \qquad [3.3]$$

Here J_x is the x-component of the **current density or current per unit cross-section perpendicular to the current flow**. In three dimensions we have the vector equation

$$\tilde{J} = \sigma\tilde{E}. \qquad [3.4]$$

This useful relation links the current density at a given location to the prevailing average electric field acting within a conductor at the same location. Because it may be derived directly from equation [3.1], equation [3.4] is often taken as an alternative statement of Ohm's law.

We previously deduced in equation [1.8] that when a charge q moves through a voltage drop of dV, the work done by the electric field was $+qdV$ joules. In a conductor this energy is dissipated as heat. The current I is defined as the charge passing through a given cross-section of the conductor every second, so that the rate of energy dissipation H is given by IdV. The voltage drop dV is IdR when a current I flows through a short length of conductor with a resistance dR, so that the rate of energy dissipated as heat each second by a current I in the short length of conductor of resistance dR is given by H in the equation

$$H = IdV = I(IdR) = I^2dR. \qquad [3.5]$$

3.2   Ionic diffusion

A molecule or ion in an aqueous solution is constantly buffeted by the surrounding water molecules. The average time that elapses between such collisions in an aqueous

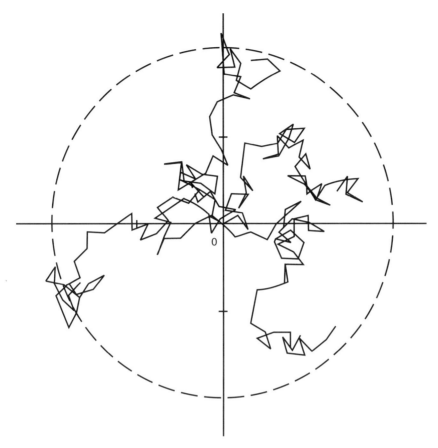

Fig. 3.2   A computer simulation of diffusion in a single plane showing four paths each consisting of 50 steps of a fixed length but in random directions. The paths all start from the origin and the predicted average excursion from the origin is shown as a dashed circle.

solution is less than 10^{-10} s at room temperature, and the average displacement between collisions is much less that the diameter of a water molecule. Under such a bombardment the molecule executes **a random walk** with every tiny step being **random in direction and length**. As an illustration, we show in Fig. 3.2 a random walk in two dimensions. This computer-generated figure shows four paths, each the result of taking 50 steps of constant length but in random directions. The broken circle shows the predicted average excursion from the origin for this random walk. The random walk of a molecule in solution differs from that in the figure in that both the direction and the length of each step vary in a random fashion.

If all of the steps were in the same direction we would expect L, the total distance covered, to grow linearly with the elapsed time t. However, for a random walk in three dimensions, the average distance traversed from its starting point in metres in

Table 3.1 Some properties of common ions important in biology. The bare (unhydrated) radius is given in units of 10^{-9} m and the given diffusion constant D in water at 25 °C must be multiplied by 10^{-9} to yield these constants in units of $m^2\,s^{-1}$.

Ion	Bare radius [10^{-9} m]	Diffusion constant D in water at 25 °C [$10^{-9}\,m^2\,s^{-1}$]
Li^+	0.060	1.03
Na^+	0.095	1.33
K^+	0.133	1.96
Rb^+	0.148	2.07
Cs^+	0.169	2.06
F^-	0.136	1.46
Cl^-	0.181	2.03
Br^-	0.195	2.08
I^-	0.216	2.04
Mg^{++}	0.065	0.071
Ca^{++}	0.990	0.079

Radii from Linus Pauling, *The Nature of the Chemical Bond.*

a time t seconds is given by L in the expression below. The distance L grows only with the square root of the elapsed time t:

$$L = \sqrt{6Dt}. \tag{3.6}$$

The constant D is called **the diffusion constant** and it is characteristic of a particular type of molecule in a particular solvent at a particular temperature. As can be seen in equation [3.6], the diffusion constant has dimensions of square metres per second. Ion diffusion is important in the operation of living systems, and the diffusion constants for some common small ions in dilute aqueous solution are listed in Table 3.1. A puzzling feature of this table is that some of the smallest univalent ions have the smallest diffusion constants, whereas it is logical to think that the smaller the ion the bigger would be its diffusion constant. The explanation is that the smaller the ion of a given valence, the bigger is the electric field at its surface and the more tightly are bound neighbouring water molecules because of their electric dipole moment. Thus the size of the ion in aqueous solution, including bound water molecules, is not simply related to its bare radius and the water-bound size may even vary inversely with the bare radius. We will return to the role of neighbouring water molecules with their electric dipole moments in Chapter 5.

Because L in equation [3.6] only grows with the square root of t, diffusion is fast over small distances but slow over larger distances. As an example, as seen from the table, the diffusion constant for the univalent potassium ion in dilute aqueous solution at 25 °C is about $1.96 \times 10^{-9}\,m^2\,s^{-1}$. To diffuse an average distance of 1 micrometre,

1 millimetre and 1 metre, the ion would require average times of 8.5×10^{-5} s, 85 s and 8.5×10^7 s or 2.7 years. It is therefore clear that as a method of signalling in a biological context, diffusion is adequate only over very small distances. Over much longer distances it becomes too slow and specialist mechanisms such as nerve conduction must be employed.

3.3   Fick's first law of diffusion

In the discussion above we were concerned with the diffusion of a single ion in a dilute solution. Here we turn our attention to the collective motions of a one type of ion in a dilute solution, particularly when there exist differences in the concentration of that type of ion in different locations. We represent the concentration of a particular type of molecule at a position defined by the coordinates (x, y, z) as $c(x, y, z)$, where the concentration c is defined as the number of this type of molecule per unit volume at that position. Note that c is **not** the molar concentration. It is found experimentally that when $c(x, y, z)$ varies with position, collective diffusive flows occur within the solution for that type of molecule. We will consider here the simple case when the concentration varies with x but does not vary with y or z so that the concentration is constant in any (y, z) plane. We will assume that over a small distance dx the concentration rises by $dc(x, y, z)$ so that the gradient of the concentration in the x-direction is given by $dc(x, y, z)/dx$. If we represent the diffusive flux in the x-direction as $j_x(x, y, z)$ and define it as the number of this type of molecule that cross a unit area perpendicular to the x-direction each second at the position (x, y, z), then the experimentally determined result, which is called **Fick's first law**, may be written as

$$j_x(x, y, z) = -D \frac{dc(x, y, z)}{dx}. \qquad [3.7]$$

The constant D is the same diffusion constant that we introduced in equation [3.6]. Thus there exists a spontaneous flux of a particular type of molecule in a particular direction when the concentration of that type of molecule falls when moving in that direction. The molecular flux is directly proportional to the negative gradient of the concentration and the constant of proportionality is the diffusion constant D, which is listed in Table 3.1 for some common ions.

We have written the diffusive flux of molecules in the x-direction as j_x and the current density in the same direction as J_x. If the diffusing molecule is charged and has a valence z then the two parameters are connected through the simple relation

$$J_x(x, y, z) = zq j_x(x, y, z), \qquad [3.8]$$

Where q is the proton charge.

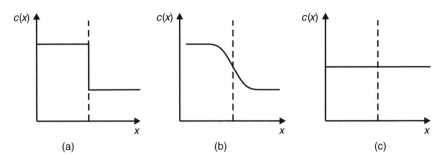

Fig. 3.3   (a) The initial concentration $c(x)$ plotted against x of a given type of particle in a solution contained in a vessel with an impenetrable barrier which prevents diffusion in the x-direction. The initial concentration is higher on the left than on the right. (b) When the barrier is removed without creating convection, there exists a concentration gradient where the barrier was and the concentration of the particle on the left decreases and on the right increases. (c) Finally, after the passage of sufficient time the concentration of the particle is constant for all values of x.

This behaviour may be illustrated by a container of aqueous solution with an impermeable partition such that the concentration of a given type of molecule is greater on one side of the container than on the other. In Fig. 3.3a, the concentration of the molecule $c(x)$ is plotted against x, where the x-direction is taken as the direction perpendicular to the partition. If the partition is removed carefully so as to cause no turbulence, a diffusive flow of the particular type of molecule occurs in the positive x-direction. As time goes on the steepness of the concentration gradient decreases and the width of the transition region increases. A graph of the concentration plotted against x is shown at some intermediate time in Fig. 3.3b. After a sufficiently long time the concentration of the molecule is constant throughout the vessel, as sketched in Fig. 3.3c, and the redistribution of this type of molecule ceases.

It is important to understand that Fick's first law describes the behaviour of individual molecules each undergoing an **independent** random walk and that it assumes **no interaction between these molecules**. In a typical biological solution the maximum concentration of a particular type of ion may be about 0.1 moles per litre, whereas the concentration of water in a dilute solution is 55.6 moles per litre so that there are over 500 water molecules for every ion of that type and direct interaction between ions of the same type will be rare. Some physical insight into the operation of Fick's first law may be had by imagining a small plane area perpendicular to the concentration gradient. Let us assume that the concentration is higher at lower values of x and decreases with x, as in Fig. 3.3b. If we imagine two equal volumes each touching the small plane area then there will be more ions of the given type in the volume at lower x than in the same volume at higher x. If all the ions execute a random walk we would expect more ions to pass through the area towards higher x from the small volume at lower x than will pass through the area towards lower x from the same volume at higher x, simply because of the greater number of ions in the volume at lower x. This is the basis for Fick's law.

Because we are dealing with the individual random walks of individual ions we would expect the flux of one type of ion to be independent of the concentration of any other type of ion, and this is found to be so in dilute solutions. Note also that the flux of a given type of ion is predicted to be dependent upon the concentration gradient but it does not depend upon the absolute value of the concentration. This is also found to be true with dilute, and therefore ideal, solutions.

3.4   Ionic motion in a constant electric field

Here we will be concerned with the effect upon an ion in a dilute aqueous solution when it is influenced by a constant electric field E_x acting in the x-direction. If this type of ion bears a charge $+q$ it will experience a constant force $F_x = qE_x$ in the x-direction. The ion is still subjected to the continuous buffeting of the neighbouring water molecules, and thus executes a random walk like every other molecule in the solution, but, as we shall see, superimposed upon this motion there is now a drift at a constant rate in the direction of the applied force. A **free** particle subject to a constant force **accelerates** at a constant rate so that its speed increases linearly with time. However, as we have discussed above, ions in aqueous solution are not free and any energy transmitted to the ion by a constant force is continuously dissipated by collisions with neighbouring molecules. The result in these circumstances is that **the response to a constant force is a constant speed** and not a constant acceleration. Such motion, dominated by dissipative viscous forces, rather than the inertial forces that determine the motion of a free particle, is characterized by an **absolute mobility**, M. The mobility is defined by the relation between the amplitude of the applied force F_x and the average drift velocity $v_D(F_x)$ in the direction of the applied field, as in the equation

$$v_D(F_x) = MF_x. \qquad [3.9]$$

In his famous paper on Brownian motion, Einstein was able to prove that the absolute mobility M and the diffusion constant D that characterize a given type of ion are related such that $D = kTM$, where k is the Boltzmann constant ($1.38 \times 10^{-23}\,\mathrm{J\,K^{-1}}$) and T is the absolute temperature. Substituting this relation and the fact that the force F_x is given by qE_x into equation [3.8], we obtain the following expression for the average drift velocity:

$$v_D(E_x) = \left(\frac{D}{kT}\right) qE_x. \qquad [3.10]$$

Thus, for example, the drift velocity of a univalent potassium cation in aqueous solution under the influence of an applied electric field of 100 V m^{-1} at 25 °C becomes 7.62 μm s^{-1}.

We can obtain the ion flux j_x or the current density J_x from equation [3.10] as

$$j_x(x, y, z) = c(x, y, z)v_D(E_x), \quad J_x(x, y, z) = qc(x, y, z)v_D(E_x), \qquad [3.11]$$

where $c(x, y, z)$ is the ionic concentration in ions per cubic metre at the position (x, y, z). Comparing equations [3.4], [3.10] and [3.11], we may deduce a value for the contribution to the conductivity σ of the solution which contains a fixed concentration c for the particular ion we are discussing:

$$\sigma = q^2 c \left(\frac{D}{kT} \right). \qquad [3.12]$$

For a dilute solution of many different ions we may simply add the contributions to the conductivity of the different types of ions with the appropriate values of q, c and D. Note that the conductivity does not depend upon the sign of q, as is consistent with the definition of the current I in the first paragraph of this chapter.

Let us now compute the motion of a potassium cation under the combined effects of thermally activated random-walk diffusion and a drift velocity caused by an applied electric field E_x. Let the ion start at the origin at a time $t = 0$. After the elapse of a time t the distance travelled in the x-direction under the influence of the electric field E_x is $v_D(E_x)t$, with $v_D(E_x)$ given by equation [3.7]. In the same time, we see from equation [3.6] that the ion will undergo a random walk with an average excursion of $\sqrt{6Dt}$ in a random direction. To combine these motions we construct a sphere of radius $\sqrt{6Dt}$ centred on the position on the x-axis such that $x = v_D(E_x)t$ and the path of the ion will start at the origin and it will end on average near the surface of the sphere.

The combination of a random walk and a constant drift is illustrated in Fig. 3.4 in two dimensions. The four computer-generated paths shown in the figure each have 50 steps. Each step is the combination of a step of fixed length but random direction and a fixed drift to the right. The broken circle shows the predicted end points of such paths on average. As in Fig. 3.2, the random paths shown have a constant step length for clarity of illustration, whereas the steps taken by a molecule in solution have a random distribution in both direction and length.

As the random walk excursion only grows as the square root of t, while the motion due to the field is directly proportional to t, it is clear that the direction of the combined motion will grow closer to the direction of the electric field as the time increases. However, for the length of displacement due to the electric field to become equal to the random excursion in a short time or over a small distance requires large applied fields. For the example we have used so far of a potassium ion in aqueous solution at 25 °C and

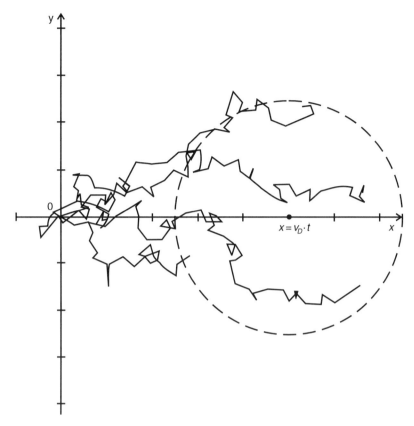

Fig. 3.4 A computer simulation of diffusion of a particular type of charged particle in a single plane when the particles are subject to thermally activated collision and the presence of an electric field acting in the *x*-direction. Each of the four paths shown consists of 50 steps. The length and direction of each step are determined by compounding the drift in the *x*-direction due to the electric field and a step of fixed length but random direction caused by the collisions. All the paths start at the origin and the dashed circle shows the predicted time-averaged termination positions of the paths.

subject to an electric field of $100\ \mathrm{V\,m^{-1}}$, the excursion of the random walk becomes equal to the distance travelled under the influence of the electric field only after an elapsed time of 202 seconds when both distances are equal to 1.54 mm. For both excursions to be equal to $10\ \mu\mathrm{m}$, taken to represent the dimensions of a cell, requires the application of a large electric field of about $15.4 \times 10^3\ \mathrm{V\,m^{-1}}$ for a period of about 8.5 ms.

3.5 Ionic motion with both an electric field and a concentration gradient

Let the concentration of a given type of ion be $c(x, y, z)$ at a position in space with coordinates (x, y, z). The negative concentration gradient in the *x*-direction is then

$-dc(x, y, z)/dx$. We may combine equations [3.7] and [3.11] to compute the total flux $j_x(x, y, z)$ of that type of ion at that point in space and in the x-direction, under the influence of this concentration gradient and an electric field with an x-component of E_x:

$$j_x(x, y, z) = -D\frac{dc(x, y, z)}{dx} + c(x, y, z)D\left(\frac{qE_x}{kT}\right). \qquad [3.13]$$

By using the facts that

$$E_x = -\frac{dV(x, y, z)}{dx} \quad \text{and} \quad \frac{d\ln[c(x, y, z)]}{dx} = \left(\frac{1}{c(x, y, z)}\right)\frac{dc(x, y, z)}{dx},$$

where $\ln[c(x, y, z)]$ is the logarithm of $c(x, y, z)$ to the base e, we arrive after some algebra at the usual expression for this total ion flux:

$$j_x(x, y, z) = -c(x, y, z)\left(\frac{D}{kT}\right)\frac{d}{dx}\{kT\ln[c(x, y, z)] + qV(x, y, z)\}. \qquad [3.14]$$

Readers familiar with thermodynamics (see Appendix 3) will recognize that the differential in equation [3.14] is the gradient with respect to x of the electrochemical potential $\mu(x, y, z)$ of the ion at a position (x, y, z), where

$$\mu(x, y, z) = \mu_0(T, p) + kT\ln[c(x, y, z)] + qV(x, y, z). \qquad [3.15]$$

Remembering that the force on an ion is the negative gradient of its electrochemical potential (see Appendix 3) and that D/kT is equal to the absolute mobility M, we could obtain equation [3.14] directly using a thermodynamic argument as follows:

$$j_x(x, y, x) = c(x, y, z) \quad \text{(drift velocity in the x-direction)}$$
$$= c(x, y, z)M \quad \text{(force on each of the ions in the x-direction)}$$
$$= c(x, y, z)\left(\frac{D}{kT}\right)\left(\frac{-d\mu(x, y, z)}{dx}\right), \qquad [3.16]$$

in agreement with equation [3.14].

Topics covered in Chapter 3

1. When a steady current flows in a conductor, Ohm's law states that the voltage drop along a short length of the conductor is proportional to the current flowing through the conductor and the resistance of the short length. The resistance is a parameter characteristic of the shape of the conductor, its internal structure and the temperature.

2. If a conducting fluid containing mobile ions is exposed to an electric field, the mobile ions in the fluid are continuously being accelerated by the internal electric field that propels them along the conductor and then colliding with the surrounding molecules such that any kinetic energy gained from the acceleration is rapidly dissipated as heat in the collisions. The result of such viscous flow is that the ions move with a constant time-averaged drift velocity in the applied field rather than accelerating in the field like free particles.

3. Ions in aqueous solution in the absence of an applied electric field are continuously colliding with the surrounding water molecules and, as a result, execute a random walk in which each step is random in both length and direction. The distance between starting and finishing points grows with the square root of the time so that small distances are covered quickly but longer distances can take long times to cover. The process is called diffusion.

4. Small ions diffusing in aqueous solution may carry with them water molecules, attached to the surface of the ion by the electrostatic attraction between the ionic charge and the permanent electric dipole moment of the water molecule.

5. When a concentration gradient exists in a solution in a given direction for a given type of ion, these ions tend to flow down the concentration gradient. The experimental finding is called Fick's first law. The flow does not depend upon any interaction between the flowing ions but is predicted theoretically for ions which each perform independent random walks.

6. The motion of a cation in aqueous solution which is exposed to an externally applied electric field is a complicated superposition of the thermally activated random walk and a steady drift along the direction of the applied field. Anions in the same circumstances perform a similar random walk but drift in the opposite direction.

Important equations of Chapter 3

1. Ohm's law connects the voltage difference ΔV applied across the ends of a conductor with the consequent current flow I by the resistance R, such that

$$\Delta V = RI.$$

The resistance depends on the shape of the conductor, the material from which it is formed and the temperature. A conductor of length L and cross-sectional area A which carries a current parallel to its length has a resistance given by

$$R = \rho \left(\frac{L}{A} \right),$$

where ρ is the resistivity of the material, defined as the resistance across opposite faces of a unit cube of the material.

2. The current density $\tilde{J}(x, y, z)$, or current per unit cross-section perpendicular to the direction of the flow, at a given position within a conductor is connected to the electric field $\tilde{E}(x, y, z)$ acting at that point by the equation

$$\tilde{J}(x, y, z) = \sigma \tilde{E}(x, y, z).$$

where σ is the conductivity of the material at that point. The conductivity is defined by the equation

$$\sigma = \frac{1}{\rho}.$$

3. Molecules or ions in aqueous solution are constantly in collision with surrounding water molecules and execute a random walk of very small steps of random lengths in random directions. The total average linear distance travelled L from its starting point at time $t = 0$ during an elapsed time t is given by

$$L = \sqrt{6Dt}.$$

The constant D is called the diffusion constant for that molecule and characterizes its diffusion in that solution at a particular temperature.

4. Fick's first law relates the diffusive flux $j_x(x, y, z)$ of a given type of molecule in the x-direction at the position (x, y, z) with the gradient of the concentration per unit volume of that molecule in the x-direction, $dc(x, y, z)/dx$:

$$j_x(x, y, z) = -D\left[\frac{dc(x, y, z)}{dx}\right].$$

5. If the molecule bears a charge q and is acted on by an electric field E_x in the x-direction, a drift in the x-direction with an average drift velocity $v_D(E_x)$ is superimposed on the random walk. The magnitude of $v_D(E_x)$ is given by

$$v_D(E_x) = D\left(\frac{qE_x}{kT}\right),$$

where k is the Boltzmann constant and T is the absolute temperature.

6. In a dilute solution the contribution to the solution conductivity made by one type of ion which bears a charge q, is at a concentration c and has a diffusion constant D, is given by

$$\sigma = q^2 c \left(\frac{D}{kT}\right).$$

7.   The flux of a given type of molecule with charge q in the x-direction under the influence of both an electric field E_x and a concentration gradient $dc(x, y, z)/dx$ in this direction is given by

$$j_x(x, y, z) = D\left[-\frac{dc(x, y, z)}{dx} + c(x, y, z)\left(\frac{qE_x}{kT}\right)\right].$$

Problems

3.1   A copper wire has a radius of 1 mm and carries a current of 1 A. If the electron density of copper is 8.45×10^{28} m^{-3} and the electron charge is -1.6×10^{-19} C, what is the mean drift velocity of the electrons along the wire?

3.2   A linear conductor with a constant cross-sectional area consists of a length L_1 of material of resistivity ρ_1 connected in series with a length L_2 of material of resistivity ρ_2. If a voltage difference V is maintained between the ends of the conductor, obtain an expression for the voltage drop across the lengths L_1 and L_2 of the two materials.

3.3   How long will it take a sodium ion on average to diffuse a distance of 10 μm from its starting point in aqueous solution at 25 °C if it has a diffusion constant $D = 1.33 \times 10^{-9}$ m^2 s^{-1}? How big an electric field must be applied if the ion is required to drift a distance of 100 μm in the direction of the electric field in the same time? Show that you would expect to find the ion within a circular cone with its apex at the starting point of the ion and with its axis parallel to the field direction. Estimate the size of the semi-angle at the apex of the cone.

3.4   Using the diffusion constants listed in Table 3.1, estimate the conductivity of a 10^{-2} molar solution of $CaCl_2$ at 25 °C.

3.5   A solution of 0.001 molar NaCl at 25 °C has a pH of 7 so that the H$^+$ and OH$^-$ ions each have a concentration of 10^{-7} molar. An electric field of 100 V m^{-1} is present in the solution. Calculate the total current in amperes that flows in the direction of the field through a unit area perpendicular to that direction carried by Na$^+$, Cl$^-$, H$^+$ and OH$^-$ ions. The diffusion constants for these ions in units of square metres per second at 25 °C are 1.33×10^{-9}, 2.03×10^{-9}, 9.31×10^{-9} and 5.30×10^{-9} respectively.

3.6   Myoglobin has a gram molecular weight of about 17 000, a density of 1.35 g cm^{-3} and a diffusion constant in aqueous solution D of about 1.1×10^{-10} m^2 s^{-1} at 20 °C. The gravitational force F_G acting on a molecule of mass m and density d in a solvent of density d_0 is given by

$$F_G = mg\left(1 - \frac{d_0}{d}\right),$$

where $g = 9.81$ m s^{-2} is the acceleration due to gravity. Using equation [3.9] and the relationship between the mobility M and the diffusion constant D noted under it, calculate the acceleration that must be applied to an aqueous solution of myoglobin in a centrifuge if the molecule is to have a drift velocity through the solution of 1 mm per hour. Express the acceleration as a multiple of the gravitational acceleration.

3.7   A linear conductor has a constant cross-section of arbitrary shape and is composed of material with a constant resistivity. A fixed voltage is applied across the ends of the conductor. Show that in the steady state (a) the internal electric field that crosses any part of a perpendicular cross-section is constant and (b) the internal electric field must be constant at any point along the length of the conductor.

4. Properties of the electric dipole and the application of Gauss's law when dielectrics are present

Most solids, liquids and gases are electrically polarizable in that, when placed in an electric field, electric charges are induced within the material and on its surface. To make physically realistic electrical predictions in the presence of such material, the origin of such induced charges must be understood and the electric fields they produce must be accounted for. In this chapter the origin and distribution of the induced charges are discussed and it is shown that the by formulating Gauss's law in terms of a new electric field vector the presence of electrically polarizable material is simply dealt with. The properties of the new field vector are discussed and compared with those of the electric field. The continuity of the new vector and the electric field vector between adjacent blocks of different polarizable material are also discussed.

4.1 The electric dipole

So far we have discussed the interaction of electric fields with charges and conductors in a vacuum. If matter is present, the electric field may induce a spatial separation within the matter of charges of different sign, and this will give rise to extra electric fields which add to the original fields. At first sight the extra induced electric fields add greatly to the difficulty of understanding electrostatic problems but, as we shall see, by introducing a new type of field we are able to include the influence of the additional induced electric fields in a simple manner. As an example of the charge separation that can occur in matter, we consider an atom or molecule placed in an electric field. The nuclei will experience a force in one direction and the electrons, bearing a charge of opposite sign, will experience forces in the opposite direction. This will result in a very small separation of the centre of gravity of the positive and negative charge distributions in the direction of the field. This situation, in which a positive charge and an equal negative charge are placed very close together, is described as an **electric dipole**. Imagine a positive charge $+q$ and a negative charge $-q$ placed a small distance d apart. Then imagine that the distance d is progressively decreased and the charges q are simultaneously increased so that the product $p = q \cdot d$ remains constant. In the limit as the distance d tends to zero this charge distribution is defined to have an electric dipole moment of p in the direction of d moving from the negative charge to the positive charge. Note that in the limit an electric dipole has a direction so that $\tilde{p}$ is a vector but, like a single charge, it is located at a point and has no finite space dimensions. The electric field of an electric dipole is sketched in Figure 4.1a. Unlike a positive charge from which the electric field radiates away, or a negative charge towards which the electric field converges, the electric field of an electric dipole leaves and returns in closed loops. If we construct a spherical Gauss law surface around the dipole we may deduce that, because the dipole has equal positive and negative charges, the total flux of the electric field through the surface is always zero no matter what the radius of the Gauss law surface. Thus all the electric field that leaves the dipole returns to it.

To reflect the axial nature of the electric dipole, the best coordinates in which to express its electric field or voltage are the two-dimensional spherical polar coordinates R and θ shown in Fig. 4.1b. The figure shows a plane surface that contains the direction of the dipole $\tilde{p}$ where R is the distance from the dipole to the field point $P(R, \theta)$ and θ is the angle between the positive direction of the dipole, which is the z-axis in the diagram, and the direction required to pass through the field point as shown in the figure. $E_R(R, \theta)$ and $E_\theta(R, \theta)$ are the two perpendicular components of the electric

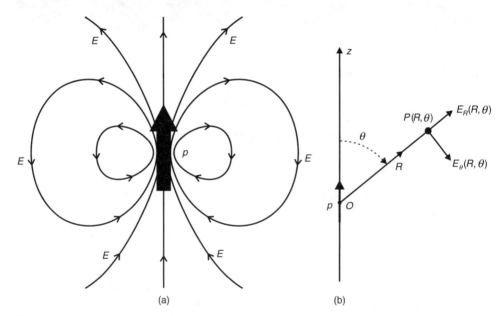

(a)   (b)

Fig. 4.1   (a) The electric field $\tilde{E}$ produced by an electric dipole $\tilde{p}$. (b) The two-dimensional polar coordinates suitable for describing the electric field created by an electric dipole. The plane of the diagram contains the field point $P(R, \theta)$ and the direction of the dipole $\tilde{p}$. The distance R is the distance from the dipole to the field point and the angle θ is the angle between the direction of the dipole and the direction of $\tilde{R}$. The two components of the electric field at $P(R, \theta)$ are shown as $\tilde{E}_R(R, \theta)$ and $\tilde{E}_\theta(R, \theta)$.

field acting at the field point P. Because of the axial symmetry of the field pattern around the dipole direction, there are no components of the field perpendicular to the plane of the figure. To obtain the full three-dimensional pattern, Fig. 4.1a or 4.1b may be rotated around the axis of the dipole. Using the expressions for the voltage around single charges (equation [1.13]) and then taking the limit described above as the distance between the charges decreases to zero to form a dipole, it can be shown that the voltage at the field point $P(R, \theta)$ due to a dipole with a dipole moment p is given by

$$V(R, \theta) = \frac{p \cos(\theta)}{4\pi\varepsilon_0 R^2}.$$

[4.1]

The polar coordinates R and θ are defined in Fig. 4.1b.

Box 4.1   Two-dimensional polar coordinates

Cartesian coordinates, which are shown in Fig. B4.1a on the left, are the simplest to understand. In two dimensions any point in the plane $P(x, y)$ may be specified by giving two coordinates x and y. The increments in these coordinates are dx and dy so that the increment of area dA, shown shaded in the figure, has a magnitude of $dxdy$. To calculate the area A of a rectangle which stretches between $x = 0$ and $x = a$ and between $y = 0$ and $y = b$ the element of area $dxdy$ is summed over these limits in a double integral:

$$A = \int_x \int_y dA = \int_{x=0}^{x=a} \int_{y=0}^{y=b} dxdy = \int_{x=0}^{x=a} dx \int_{y=0}^{y=b} dy = ab.$$

The rates of change of the coordinates x and y with time, are simply dx/dt and dy/dt and these are used to represent velocities.

Although familiar and simple, Cartesian coordinates are of little use when discussing natural objects because these very seldom have rectangular symmetry. Of much greater value are polar coordinates which are suitable to describe natural objects that approximate in shape to circles, ellipses, spheres and cylinders. Figure B4.1b shows two-dimensional polar coordinates. Here a point in space $P(R, \theta)$ is specified by giving the distance R of the point from the origin 0 and the angle θ that the vector $\tilde{R}$ makes with the z-axis. The general recipe for generating the increments in any coordinate system is to let all the other coordinates remain fixed while one coordinate is allowed to increase slightly. Keeping θ constant and allowing R to increase slightly gives one increment dR as in the figure. Keeping R constant and allowing θ to increase slightly results in a movement of the field point $P(R, \theta)$ round the circumference of a circle of radius R, moving a small distance of length $Rd\theta$ as

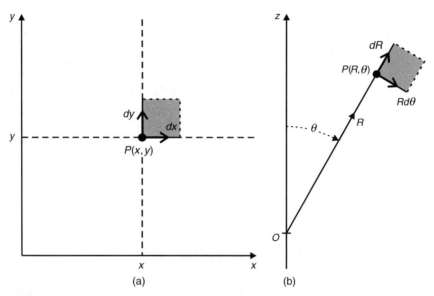

Fig. B4.1 (a) Two-dimensional Cartesian coordinates where a position $P(x, y)$ in the plane is specified by giving the distances x and y along the horizontal and vertical axes respectively. The increments in distance are dx and dy and they are perpendicular to each other, so that the increment in area is their product $dxdy$. (b) Two-dimensional polar coordinates where the position of the point $P(R, \theta)$ is specified by giving its distance from the origin R such that the angle between the direction of $\tilde{R}$ and that of the z-axis is given by θ. One length increment of length dR is obtained by allowing R to increase slightly while keeping θ constant. The other length increment is obtained by keeping R constant and allowing θ to increase slightly. The length of this increment is $Rd\theta$ and it is perpendicular to dR. The increment in area is thus $dRRd\theta = RdRd\theta$.

shown in the figure. This last result is a direct consequence of the fundamental definition of the size of an angle $d\theta$ in radians subtended at the centre of a circle of radius R by a length dL of the circumference as

$$d\theta = \frac{dL}{R} \quad \text{or} \quad dL = Rd\theta.$$

Thus the element of area dA in the two-dimensional polar coordinates, shown shaded in the figure, has a magnitude of $dRRd\theta$ or $RdRd\theta$. To calculate the area A of a plane annular ring with inner radius a and outer radius b, the elementary area is integrated between the limits $a \le R \le b$ and $0 \le \theta \le 2\pi$:

$$A = \int_R \int_\theta dA = \int_{R=a}^{R=b} \int_{\theta=0}^{\theta=2\pi} RdRd\theta$$

$$= \int_{R=a}^{R=b} RdR \int_{\theta=0}^{\theta=2\pi} d\theta = \left[\frac{b^2}{2} - \frac{a^2}{2}\right] 2\pi = \pi(b^2 - a^2).$$

The velocities (v_R, v_θ) in the directions of the distance increments dR and $Rd\theta$, are obtained by dividing these increments by dt and taking the limit as dt tends to zero to yield differentials, so that we obtain the expressions

$$v_R = \frac{dR}{dt} \quad \text{and} \quad v_\theta = R\left(\frac{d\theta}{dt}\right).$$

When the geometry studied has axial symmetry, such as a representation of the field of an electric dipole in Fig. 4.1a, the full three-dimensional picture may be obtained by a rotation of a two-dimensional polar coordinate plot, like that shown above, about the z-axis.

To obtain the electric field components from an expression for the voltage we may use the fact, expressed in equation [1.8], that the change in voltage dV experienced during a small movement dL is given by $-E_L dL$, where E_L is the component of the electric field along the direction of dL. Using polar co-ordinates (see Box 4.1), the differential increment in the direction of R is obtained by keeping θ constant and slightly increasing R to $R + dR$. If we move this small increment dR in the direction of R, the change in the voltage is given by $dV(R, \theta) = -E_R(R, \theta)dR$. Thus, by differentiation with respect to R of the expression for $V(R, \theta)$, we have

$$E_R(R, \theta) = -\frac{dV(R, \theta)}{dR} = +\frac{2p\cos(\theta)}{4\pi\varepsilon_0 R^3}. \qquad [4.2]$$

Similarly, the differential increment in the θ-direction, shown as the direction of E_θ in Fig. 4.1b, is obtained by keeping R constant and allowing θ to increase a small amount to $\theta + d\theta$. In such a change the point P moves along the circumference of a circle of radius R when the radius rotates through a small angle $d\theta$. The length of the arc traced out by the field point P is $Rd\theta$ and its direction is perpendicular to R. Thus the change in voltage $dV(R, \theta)$ during this movement is given by $dV(R, \theta) = -E_\theta(R, \theta)Rd\theta$. Thus we may deduce that

$$E_\theta(R, \theta) = -\left(\frac{1}{R}\right)\frac{dV(R, \theta)}{d\theta} = \frac{p\sin(\theta)}{4\pi\varepsilon_0 R^3}. \qquad [4.3]$$

Notice again that at a field point $P(R, \theta)$ in a plane that contains the dipole such as that shown in Fig. 4.1b, the electric field is directed within this plane and there are no components of the field at $P(R, \theta)$ perpendicular to this plane. This is intuitively obvious as the two charges of opposite sign that may be considered to form the dipole lie within the plane and produce at $P(R, \theta)$ electric fields within this plane. The full three-dimensional picture of the electric fields surrounding an electric dipole may be obtained by rotating a planar picture such as that in Fig. 4.1b around the direction of the dipole. We note that whereas the voltage and electric field around a single charge fall off as R^{-1} and R^{-2} respectively, the voltage and electric field of an electric dipole fall off as R^{-2} and R^{-3} respectively. The faster fall-off with R for the dipole is to be expected because of the partial cancellation of the fields and voltages of the two closely spaced charges of opposite sign.

To obtain the total electric field in a general direction, the two components of the electric field, E_R and E_θ, in the chosen direction and at the field point $P(R, \theta)$ may be compounded in the usual way. For example, if we wish to know the magnitude of the electric field in the $+z$-direction at $P(R, \theta)$ we may deduce it as follows:

$$E_Z(R, \theta) = E_R \cos(\theta) - E_\theta \sin(\theta) = \left(\frac{p}{4\pi\varepsilon_0 R^3}\right)[2\cos^2(\theta) - \sin^2(\theta)]$$

$$= \frac{p}{4\pi\varepsilon_0 R^3}[3\cos^2(\theta) - 1], \qquad\qquad [4.4]$$

using the fact that $\sin^2(\theta) + \cos^2(\theta) = 1$.

4.2   The interaction of electric dipoles with electric fields

We will start with the simplified model of an electric dipole as a pair of charges $+q$ and $-q$ separated by a small distance d. If we place this model in a uniform electric field $\tilde{E}$ as in Fig. 4.2 which is in the plane containing the two charges, the two charges,

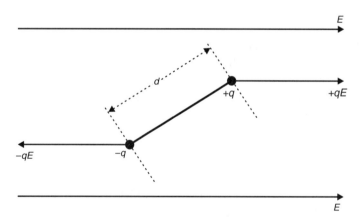

Fig. 4.2   A model of an electric dipole consisting of charges $+q$ and $-q$ separated by a distance d, situated in a uniform electric field $\tilde{E}$. The equal in magnitude but oppositely directed forces $+qE$ and $-qE$ acting on the two charges are shown. Because the forces are equal but oppositely directed there is no net translational force on the model dipole but there does exist a couple tending to twist the dipole about an axis perpendicular to its direction.

experience equal and opposite forces of magnitude $+qE$ and $-qE$ along the field direction. The two forces acting on the dipole are equal in amplitude but opposite in direction, so that we would not expect the model dipole to experience any translational force in a uniform electric field. This conclusion remains true as we take the limit as d tends to zero while the product $qd = p$ remains constant to form a real dipole, so we may be sure that **an electric dipole in a uniform electric field experiences no translational force**.

The two forces that act upon the model dipole in the electric field are equal in magnitude but act in opposite directions along parallel lines and constitute a **couple** $C(\theta)$ that acts upon the model dipole. The strength of a couple (see Appendix 1), which measures its ability to exert a twist, is defined as the amplitude of one force multiplied by the perpendicular distance between the two forces. In this case each force has a magnitude of qE and the perpendicular distance between the two forces is $d\sin(\theta)$, where θ is the angle between the direction of the dipole moment $\tilde{p}$ and the electric field $\tilde{E}$. When we take the limit as d tends to zero while the product $qd = p$ remains fixed, so as to convert the model dipole to a real dipole, we find that the amplitude of the couple $C(\theta)$ remains constant and is given by

$$C(\theta) = qEd\sin(\theta) = qdE\sin(\theta) = pE\sin(\theta). \qquad [4.4]$$

From Fig. 4.2 it can be seen that the twisting action of the couple acts along a direction perpendicular to the directions of the dipole and of the electric field in the direction of advance of a right-handed thread rotated from the direction of the dipole to that of the field. Thus the couple always acts such as to reduce the angle θ and align the direction of the dipole moment $\tilde{p}$ with the direction of the electric field $\tilde{E}$. The couple is zero when the dipole moment is aligned ($\theta = 0$) with the applied field, and is a maximum when the dipole moment is at right angles to the field ($\theta = 90°$).

We will now calculate the potential energy of an electric dipole $\tilde{p}$ which it has because of its orientation at an angle θ to an electric field $\tilde{E}$. Let us take the model dipole shown in Fig. 4.2 and take as the x-direction the direction of the electric field so that $E = -dV/dx$. The energy of the positive charge in the electric field is $+qV_+$, where V_+ is the voltage at the position of the positive charge. Similarly, the energy of the negative charge is $-qV_-$, so that the energy of both charges is given by $U_p(\theta) = q(V_+ - V_-)$. Now V_+ is smaller than V_- because the two charges are separated by a distance of $d\cos(\theta)$ in the direction of the electric field, and the voltage drops when moving in this direction. We can write the difference between the voltages as $V_+ - V_- = (dV/dx)[d\cos(\theta)]$, where (dV/dx) is the gradient of the voltage with distance x. Moving from the model dipole to the real dipole, we take the usual limit as d tends to zero and the product $qd = p$ remains constant to obtain the energy $U_p(\theta)$ of the electric dipole when it is oriented at an angle θ to the prevailing electric field $\tilde{E}$:

$$U_p(\theta) = q(V_+ - V_-) = qd\frac{dV}{dx}\cos(\theta) = -pE\cos(\theta). \qquad [4.5]$$

Extra energy is stored in the electric dipole when it is twisted away from a direction parallel to the prevailing electric field against the restoring forces exerted upon it by the electric field, just as energy is stored in a compressed spring by compressing it against the restoring forces exerted by the spring. The dipole has a minimum energy of $U_p(\theta = 0) = -pE$ when it is parallel to the applied electric field and a maximum energy of $U_p(\theta = 180°) = +pE$ when it is directed anti-parallel to the applied field. When $\theta = 90°$ the energy is clearly zero as the two voltages V_+ and V_- used above are equal. To twist an electric dipole from parallel to the electric field to anti-parallel requires energy of $2pE$.

4.3   Induced charges in dielectrics

A **dielectric** is a substance which, when it is placed in an electric field, becomes electrically polarized such that electric dipole moments are created within it. We mentioned above the case when the electron clouds surrounding the nuclei of a molecule become slightly distorted in the field so that they are no longer symmetrically disposed about the nuclei and an electric dipole moment is created. Another important example is a substance such as water within which the molecules possess a **permanent electric dipole moment**. Water molecules have a permanent electric dipole moment because the two positively charged protons covalently bonded to the oxygen atom are not symmetrically placed relative to the centre of gravity of the negative charge on the oxygen atom. The result is a molecular electric dipole moment in a direction bisecting the angle between the two oxygen–hydrogen bonds and directed away from the oxygen molecule. When water is placed in an electric field there is a tendency for the electric dipole moments of the water molecules to align in the prevailing electric field because of the couple exerted by the field on their dipole moments that we discussed above. This aligning tendency is opposed and disrupted by the thermally activated collisions with neighbouring water molecules, as we discussed in Chapter 3. We will soon calculate that the degree of alignment of the electric dipoles within a dielectric, created by electric fields of normal magnitude, is very small, but it has a none the less profound effect upon any electric fields created within its boundaries. Whether from distortion of the electron orbits or from the rotation of permanent electric dipole moments, an electric dipole moment $\tilde{P}$ per unit volume is induced when any dielectric substance (solid, liquid or gas) is exposed to an electric field. Note that $\tilde{P}$ is a vector quantity with both a direction and an amplitude which may change throughout the dielectric.

As in Section 2.1, where we described the electric field within a conductor, we must, in a discussion of the electric fields within a dielectric material, distinguish between **microscopic** or **local** electrical effects and their average over **macroscopic** distances. Near a polarized molecule the electric field may be very large and it will vary markedly over distances comparable to those between molecules. However, in dealing with macroscopic electric effects inside a dielectric material, we are interested in electrical quantities averaged over macroscopic distances. The average electric field $\tilde{E}$ and the electric dipole moment $\tilde{P}$ per unit volume are two such macroscopic averages. Strictly speaking, in order to obtain a typical average value, we average the electrical quantities over a macroscopic volume and then take the average of the volume average as we move the chosen volume small distances in each direction. This ensures that the volume chosen is typical of the interior of the dielectric material.

In Fig. 4.3a is sketched a cube of material with one opposite pair of plane faces perpendicular to the x-direction. The macroscopic average electric field within the dielectric material is in the x-direction and has a constant amplitude of E_x independent of the value of x. Since E_x is constant, the induced electric dipole moment per unit volume is also constant with a value P_x and is in the x-direction. Because P_x is constant there is no net charge induced within the body of the cube. In any small volume at a given value of x there are as many positive ends of the dipoles induced in the material at slightly smaller values of x as there are negative ends of the dipoles induced in the material at slightly larger values of x, and the total charge induced in the small volume is zero.

This is not true at the surface. At the right-hand face in the figure the positive ends of the induced dipoles at the surface are uncompensated, while the negative ends of these dipoles within the cube are compensated by the positive ends of dipoles induced in the material at slightly smaller values of x. The result is a layer of surface charge with a charge density (charge per unit area of the surface) of $\sigma = P_x$. In a more general geometry, in which the surface is not perpendicular to the field, the **surface charge density** or charge per unit area of the surface will be equal to the component of the dipole moment per unit volume $\tilde{P}$ normal to the surface, which we write as P_N. The component of $\tilde{P}$ parallel to the surface P_P will induce no net surface charge because no charge penetrates the surface and thus the charge cancellation between the positive and negative ends of the induced dipoles, which occurs within the body of the material, will occur for the dipoles created by P_P. In a similar manner, at the left-hand edge of the cube in Fig. 4.3a the induced surface charge per unit area will be $\sigma = -P_x$ or, in a general geometry, $\sigma = -P_N$. The total charge induced in the body must be zero as the body is isolated with no means of conveying charge to it or away

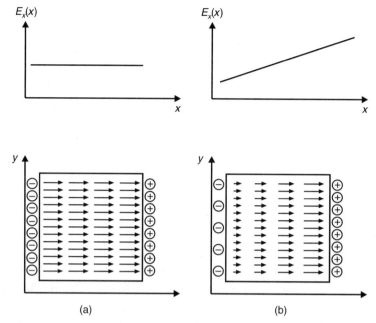

Fig. 4.3   (a) A square block of electrically polarizable material with a uniform electric field $E_x(x)$ acting within it. The electric dipole moment per unit volume induced within the material is constant. There is a positive surface charge density induced on the right-hand face of the block and an equal negative surface charge density on the left-hand face due to the uncompensated ends of the electric dipole moments induced within the material. Within the material there is no volume charge density induced due to the cancellation of charge between the positive and negative ends of the internal induced dipoles at a given position. (b) The same block with an electric field $E_x(x)$ within it that increases in magnitude as x increases. There is a positive surface charge density on the right-hand face and a smaller negative surface charge density on the left-hand face due to the uncompensated ends of the induced electric dipole moments within the material at these surfaces. Within the material there is a negative volume charge density because at any position the positive ends of the induced dipoles immediately to the left of the chosen position are smaller than the negative ends of the induced dipoles immediately to the right of the chosen position.

from it. Thus in the cube shown in the Fig. 4.3a, no body charge is induced but surface charges are induced when the local direction of $\tilde{P}$ has a component normal to the surface. The total induced charge must always sum to zero. This argument can clearly be extended to a body of any shape where the surface charge density will depend on the local value of P_N.

We now turn to Fig. 4.3b, which represents the same cube of material such that the average internal electric field E_x remains in the x-direction but has an amplitude that increases as x increases. $P_x(x)$ is now no longer constant but increases with increasing x. Using the same argument as above, we may deduce that the surface charge densities on the two faces normal to the x-axis are equal to the normal component of $\tilde{P}(x)$ but now these charge densities are different due to the variation of $\tilde{P}(x)$ with x and so become $\sigma(x_1) = -P_N(x_1)$ and $\sigma(x_2) = +P_N(x_2)$, where x_1 and x_2 are the x-coordinates of the two faces normal to the x-axis. Also because of the variation of $P(x)$ with x, there is now a body charge induced. If we consider a small volume at a position given by the coordinate x, the positive charge introduced into this volume by the positive ends of the dipoles induced at slightly smaller values of x can no longer compensate for the negative ends of the slightly stronger dipoles induced at slightly larger values of x. The induced negative charge per unit volume will be proportional to the rate at which the dipole moment per unit volume $P_x(x)$ grows with x. Thus the induced body charge per unit volume $\rho(x)$ is given by

$$\rho(x) = -\frac{dP_x(x)}{dx}. \tag{4.6}$$

Notice that although it is conventional to call both σ and ρ charge densities, they have different dimensions as ρ is the charge per unit volume and has units of coulombs per cubic metre while σ is the charge per unit area and has units of coulombs per square metre.

In a more general geometry the result for the surface charge induced per unit area is substantially unchanged but is given by $\sigma(x, y, z)$ in the expression

$$\sigma(x, y, z) = +P_N(x, y, z), \tag{4.7}$$

where $P_N(x, y, z)$ is the **outward** component of $\tilde{P}$ normal to the surface at the position given by (x, y, z). The induced body charge per unit volume $\rho(x, y, z)$ at the position (x, y, z) in a general polarization $\tilde{P}$ that varies along all three co-ordinate directions becomes the sum of three expressions like that in equation [4.6] and may be written as

$$\rho(x, y, z) = -\left[\frac{dP_x(x, y, z)}{dx} + \frac{dP_y(x, y, z)}{dy} + \frac{dP_z(x, y, z)}{dz}\right]$$
$$= -\text{div}[\tilde{P}(x, y, z)]. \tag{4.8}$$

$\text{div}[\tilde{P}(x, y, z)]$ is a scalar quantity called the **divergence of the vector** $\tilde{P}(x, y, z)$. This is a quantity that plays a role in the theory of vector fields. However, we will use it simply as a compact way of writing the sum of the three differentials shown above (see Appendix 1). It is not necessary for our purposes to understand the full implications of vector field calculus.

Once again the total charge induced over the whole isolated body by an applied electric field must be zero. The applied field merely redistributes slightly the charges within the body which had no uncompensated charge before the field was applied and thus must have no total charge in the field. In the cube shown in Fig. 4.3b the negative body charge density $\rho(x)$ just compensates for the fact that the positive surface charge density $\sigma(x_2)$ is greater than the negative surface charge density $\sigma(x_1)$. These arguments can clearly be generalized to apply to a body of any shape provided that the local values of $P_x(x, y, z)$, $P_y(x, y, z)$, $P_z(x, y, z)$ and $P_N(x, y, z)$ are employed.

4.4   Gauss's law with dielectrics present

At first sight the presence of dielectric material would seem to complicate the application of Gauss's law considerably. As before, we must take account of the initial charge distributions that define the electrical situation, but now we must also take account of charges induced on the surface and within the volume of all dielectrics that are located within the Gauss law surface. To make matters worse, the induced charges create extra electric fields which in turn induce further charges in the dielectrics and some iterative procedure might be thought necessary. The fact that we can deal with dielectrics simply when applying Gauss's law is due to a mathematics theorem called the **divergence theorem** (see Appendix 1). This purely geometrical theorem applies to any vector $\tilde{v}(x, y, z)$ and relates the volume integral of the divergence of that vector $\mathrm{div}[\tilde{v}(x, y, z)]$ over a given volume to the flux of that same vector through the surface enclosing the given volume:

$$\iiint_V \mathrm{div}[\tilde{v}(x, y, z)]dV = \iint_S v_N dS$$

or

$$\iiint_V \left[\frac{dv_x(x, y, z)}{dx} + \frac{dv_y(x, y, z)}{dy} + \frac{dv_z(x, y, z)}{dz} \right] dxdydz = \iint_S v_N dS. \qquad [4.9]$$

We must now make a distinction between the charges that are induced within or on any dielectric by an applied electric field, which we will call **bound charges** as they are bound to the atom or molecule in which they were induced, and any other charges that we distributed in setting up the problem, which we call **free charges** as we are free to distribute these as we choose. In any small volume $dxdydz$ at a position (x, y, z) we define two charge densities, the bound charge density ρ_b and the free charge density ρ_f, so that the total charge within the infinitesimal volume $dxdydz$ is given by $(\rho_b + \rho_f)dxdydz$. We will now apply Gauss's law to this situation. In the first instance we will consider a simplified geometry and not the most general geometry. We will consider the case in which the Gauss law surface is wholly contained within a block of homogeneous dielectric so that we have to take account of the induced body charge density throughout the dielectric but not the surface charge density induced on the surface of the dielectric. Gauss's law, as in equation [1.7], may be written

$$\iint_S E_N dS = \iiint_V \frac{\rho_f + \rho_b}{\varepsilon_0} dV$$

or

$$\iint_S \varepsilon_0 E_N dS = \iiint_V [\rho_f + \rho_b] dV. \qquad [4.10]$$

But we know from equation [4.8] that $\rho_b = -\text{div}[\tilde{P}(x, y, z)]$ and from the divergence theorem applied to the vector $\tilde{P}(x, y, z)$ that

$$\iiint_V \rho_b dV = -\iiint_V \text{div}[\tilde{P}(x, y, z)] dV = -\iint_S P_N dS. \qquad [4.11]$$

This equation converts a volume integral of ρ_b to a surface integral of P_N. If we apply this result to equation [4.10] we can convert the volume integral of ρ_b on the right-hand side to a surface integral which we take to the left-hand side to join the surface integral of E_N. Thus transformed, we may rewrite equation [4.10] as

$$\iint_S (\varepsilon_0 E_N + P_N) dS = \iint_S (\varepsilon_0 \tilde{E} + \tilde{P})_N dS = \iiint_V \rho_f dV. \qquad [4.12]$$

In words, this new form of **Gauss's Law states that the flux of the vector $(\varepsilon_0 \tilde{E} + \tilde{P})$ out of any closed surface is equal to the total amount of uncompensated free charge enclosed within that surface**. The new vector is called the **electrical displacement** and is written as $\tilde{D}$, so that

$$\tilde{D} = \varepsilon_0 \tilde{E} + \tilde{P}. \qquad [4.13]$$

Thus, the new form of Gauss's law, which takes account of the induced body charges in dielectrics, may be simply written as

$$\iint_S D_N dS = \iiint_V \rho_f dV. \qquad [4.14]$$

Although we will not prove it here, it can be shown that this form of Gauss's Law holds even when the Gauss law surface intersects the surfaces of dielectric bodies, so that it takes full account of the induced surface charges on dielectric bodies as well as the induced body charges within dielectrics. Equation [4.14] becomes the new form of Gauss's law which applies in all electrical situations.

The emergence of so simple a law as that in equation [4.14] that applies in the very complicated situation in which we take full account of the charges induced in and on dielectrics seems almost miraculous, but to apply the law properly requires a clear understanding of the newly defined vector $\tilde{D}$ and, in particular, its relation to the electric field $\tilde{E}$. This we tackle in the next section.

4.5   The properties of the electric displacement $\tilde{D}$

Many of the difficulties that are experienced by students learning about electricity for the first time stem from confusion between the properties of the vectors $\tilde{E}$ and $\tilde{D}$. We will describe here two of the major differences between these vectors. In our discussion of Gauss's law in the absence of dielectrics in Chapter 1 we concluded that the electric field $\tilde{E}$ originates on positive charges and terminates on negative charges and that charges are the only sources and sinks of electric fields. In an identical manner, an inspection of equation [4.14] leads to the conclusion that the only sources or sinks of the vector $\tilde{D}$ are free charges. The vector $\tilde{D}$ ignores the presence of induced or bound charge. However, induced charge is just as real physically as free charge and any distinction between these forms of charge is arbitrary and non-physical. This observation leads to the conclusion that although the invention of this new vector $\tilde{D}$ results in a particularly simple form of Gauss's law, the vector $\tilde{D}$ itself has strange and non-physical properties. A second important fact to keep in mind is that **the only real physical force vector is $\tilde{E}$**. Thus the force on a charge q is always $q\tilde{E}$ and never $q\tilde{D}$, and this is true whether there are dielectrics present or not. It is even true within a dielectric.

The situation we are faced with may be summed up as follows. The presence of induced charges in and on dielectrics makes the application of the original form of Gauss's law difficult. However, by invoking the divergence theorem and inventing a new vector $\tilde{D}$ we can derive a new and simple form of Gauss's law that holds universally and, in particular, in the presence of dielectrics. The price we have to pay for this apparent simplification is that the new vector $\tilde{D}$ has non-physical properties, and in order to extract real physical information we always need to convert a knowledge of the vector $\tilde{D}$ to a knowledge of the vector $\tilde{E}$.

In all gas and liquid dielectrics and in isotropic solid dielectrics, the induced dipole moment per unit volume $\tilde{P}$ is in the direction of the electric field $\tilde{E}$ that induces the moment. In these cases we can define a simple scalar constant to connect the amplitudes of the vectors $\tilde{D}$ and $\tilde{E}$, because $\tilde{D}$, $\tilde{E}$ and $\tilde{P}$ all point in the same direction, so that we may rewrite equation [4.13] as

$$\tilde{D} = \varepsilon_0\tilde{E} + \tilde{P} = \varepsilon_R\varepsilon_0\tilde{E},$$

[4.15]

where the constant ε_R, which characterizes the dielectric at that point, is called the **relative dielectric constant**. In cases where ε_R is known throughout the region of interest it becomes a very simple matter to deduce E when D is known. In some anisotropic crystalline solids the distortion of the electron orbits resulting from an applied electric field may depend upon the direction within the crystal in which $\tilde{E}$ is applied. In these cases $\tilde{P}$ may not be parallel to the inducing field $\tilde{E}$ so that equation [4.15] is no longer true and equation [4.13] must always be used. However, this situation is very rare and we will assume that we are dealing with isotropic solid dielectrics or with liquid or gaseous dielectrics within which $\tilde{P}$ is parallel to $\tilde{E}$ and equation [4.15] is valid.

A second useful technique for deducing $\tilde{E}$ when $\tilde{D}$ is known is a knowledge of how these two vectors change at the junction of two different dielectrics. In Fig. 4.4a is shown the plane junction between two uniform dielectrics with relative dielectric constants ε_{R1} and ε_{R2}. The component of the electric field E_{T1} which is tangential to the surface in dielectric 1 is shown, with the parallel tangential component in dielectric 2 shown as E_{T2}. We construct a small rectangular path with two sides of length a parallel to the tangential fields and two very short perpendicular closing sides of length b. We now calculate the line integral of the electric field around this path as in equation [1.12] and equate it to zero because the voltage is single-valued.

$$0 = \oint E_R dR = a(E_{T1} - E_{T2}) \quad \text{or} \quad E_{T1} = E_{T2}. \tag{4.16}$$

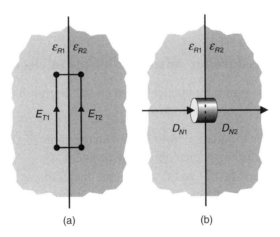

(a)   (b)

Fig. 4.4 (a) The electric field components E_{T1} and E_{T2} tangential to the plane interface between materials with relative dielectric constants ε_{R1} and ε_{R2}. By equating the line integral of the electric field $\oint E_R dR$ around a closed circuit to zero, it can be proved that $E_{T1} = E_{T2}$. (b) A Gauss law cylindrical surface embedded in the plane surface with its axis normal to the interface. If there is no free charge on the interface, applying Gauss's law in $\tilde{D}$ to the cylinder proves that the normal components of $\tilde{D}$ at either side of the interface are equal so that $D_{N1} = D_{N2}$.

When deducing this equation we employ the same arguments as those used to deduce equation [2.1]. We make the length b short to minimize these contributions to the integral and rely on the fact that when performing a complete circuit we cross the surface along the sides of length b in opposite directions relative the surface. Any contributions to the integral from the presence of the surface bound charge when we cross the interface will cancel to zero when we make the transition twice in opposite directions. **Thus we have deduced that the tangential component of $\tilde{E}$ is conserved across the boundary between two dielectrics**.

In Fig. 4.4b is shown a cylindrical Gauss law surface with its plane ends parallel to the plane interface between the two dielectrics 1 and 2. Also drawn in the figure are the two components of $\tilde{D}$ in the two dielectrics normal to the interface at that position, D_{N1} and D_{N2}. By making the cylinder arbitrarily short we can exclude the flux through the cylinder of any tangential components of $\tilde{D}$. Although there may be induced polarization charges at the interface, there is no free charge, so that the free charge included in the cylinder is zero. Applying Gauss's law as in equation [4.14] to this situation, we deduce that

$$D_{N1} = D_{N2}. \qquad [4.17]$$

Thus we have deduced that the normal component of $\tilde{D}$ is conserved across the interface between two dielectrics. This is the second boundary condition which often allows us to deduce values of $\tilde{E}$ or $\tilde{D}$ within a given dielectric when we know their values within a second dielectric and when the two dielectrics have a common interface.

4.6   General procedure for solving problems in electrostatics

A recipe that is very generally applicable when solving problems in electrostatics when dielectrics are present is as follows:

1. Use Gauss's law as in equation [4.14] to obtain the value of $\tilde{D}$ that results from the presence of the free charges that define the problem.
2. Use the relation between D and E within one dielectric as in equation [4.15] to obtain the amplitude of the electric field E, where the value of D is known.
3. Use the boundary conditions expressed in equations [4.16] and [4.17] to transform the values of E and D in one dielectric to those within other dielectrics that share a common boundary.
4. Calculate the line integral of $\tilde{E}$ between two points to deduce the change in voltage between them using equation [1.11].

Examples of the application of this recipe are given in the next chapter.

Topics covered in Chapter 4

1.   An electric dipole may be imagined to consist of a positive charge and an equal-magnitude negative charge placed very close together. The electric dipole moment has a magnitude equal to the product of the magnitude of the one charge and its separation from the other charge. The direction of an electric dipole points from the negative charge towards the positive charge.

2.   The electric field of an electric dipole consists of closed loops which originate on the positive charge and terminate on the negative charge. The amplitude of the dipolar electric field falls off as the inverse cube of the distance from the dipole.

3.   In an externally applied uniform electric field an electric dipole experiences no translational force but does experience a torque tending to align the electric dipole moment with the positive direction of the electric field.

4.   A dielectric material contains within it molecules which can distort or rotate in an externally applied electric field such that dipoles are formed within the material. As a result an electric dipole moment per unit volume is created within the material. These dipoles create extra electric fields which must be taken into account when solving electrostatic problems in the presence of dielectric material.

5.   When the electric field within a dielectric material is uniform, charges are induced on the surfaces but not within the body of homogeneous dielectric material. In a non-uniform internal electric field, charges are induced within the body of dielectric material and on its surfaces.

6.   In the absence of dielectric material, Gauss's law is stated in terms of the flux out of a closed surface of the electric field $\tilde{E}$. In the presence of the charges induced within and on the surface of dielectric material, Gauss's law may be most simply stated in terms of the flux of a new vector, the electric displacement $\tilde{D}$, out of any closed surface.

7.   The electric displacement is a vector with physically unrealistic properties. The solution of an electrostatic problem in the presence of dielectric material often starts by finding the value of the electric displacement. However, before physically realistic information can be obtained, the value of the electric field must be deduced from the known values of the electric displacement.

8.   In homogeneous solids, or in liquids and gases, the electric displacement within a dielectric material points in the direction of the local electric field and its amplitude is determined by multiplying the amplitude of the electric field by a scalar constant characteristic of the material.

9.   At the plane boundary between two homogeneous dielectrics the tangential component of the electric field and the normal component of the electric displacement are conserved.

Important equations of Chapter 4

1.   The voltage created at a field point $P(R, \theta)$ in a vacuum by an electric dipole p located at the origin of coordinates and pointing along the z-axis is given by

$$V(R, \theta) = \frac{p \cos(\theta)}{4\pi\varepsilon_0 R^2}.$$

The field point $P(R, \theta)$, expressed in polar coordinates, is located a distance R away from the dipole and such that the angle between the direction $\tilde{p}$ and the direction of R is θ.

2.   An electric dipole $\tilde{p}$ situated in a uniform electric field $\tilde{E}$ such that the angle between the directions of $\tilde{p}$ and $\tilde{E}$ is θ has an energy $U(\theta)$ and experiences a couple $C(\theta)$ such that

$$U(\theta) = -pE \cos(\theta) \quad \text{and} \quad C(\theta) = +pE \sin(\theta).$$

3.   Within a dielectric material an electric field $\tilde{E}(x, y, z)$ will induce an electric dipole moment per unit volume of $\tilde{P}(x, y, z)$ at a position (x, y, z). The charge density per unit volume $\rho(x, y, z)$ induced within the body of the dielectric and the charge density per unit area $\sigma(x, y, z)$ induced on the surface of the dielectric are given by

$$\rho(x, y, z) = -\text{div}[\tilde{P}(x, y, z)] \quad \text{and} \quad \sigma(x, y, z) = +P_N,$$

where $\text{div}[\tilde{P}(x, y, z)]$ is written in full in equation [4.8] and P_N is the outward-pointing normal component of $\tilde{P}$ at the surface of the dielectric.

4.   When dielectrics are present, Gauss's law may be written as

$$\iint_S D_N dS = \iiint_V \rho_f dV,$$

where ρ_f is the density per unit volume of the free charge at the position (x, y, z) and the vector $\tilde{D}$ is connected to the electric field $\tilde{E}$ by

$$\tilde{D} = \varepsilon_0 \tilde{E} + \tilde{P} \quad \text{or} \quad \tilde{D} = \varepsilon_R \varepsilon_0 \tilde{E}.$$

The parameter ε_R is the relative dielectric constant which characterizes the electrical polarizability of the dielectric at a particular position. The first of these equations is quite general and the second holds in liquids, gasses and homogeneous solids where the induced dipole moment per unit volume $\tilde{P}$ is parallel to $\tilde{E}$.

5.   At the plane boundary between two dielectric materials with relative dielectric constants of ε_{R1} and ε_{R2} the normal component of $\tilde{D}$, written as D_N, and the tangential component of $\tilde{E}$, written as E_T, are conserved so that the following equations hold:

$$D_{N1} = D_{N2} \quad \text{and} \quad E_{T1} = E_{T2}.$$

Problems

4.1   A charge $+Q$ is situated at a position $P(R, \theta)$ relative to an electric dipole as shown in Fig. 4.1b. If the dipole moment is p, use equation [4.1] to deduce the interaction energy of the dipole and the charge. If R remains constant, obtain values of this interaction energy when $\theta = 0$, $90°$ and $180°$. Give a physical explanation of these results in terms of the two charges that may be thought to compose the dipole.

4.2   A uniform electric field is directed along the x-axis. Show that an electric dipole experiences no couple acting on it if it is pointing along this axis or in the opposite direction. Explain why, in a sufficiently strong electric field, an electric dipole in thermal equilibrium with its surroundings is much more likely to be found parallel rather than anti-parallel to the applied field.

4.3   A water molecule has a permanent electric dipole moment of 6.17×10^{-30} C m. Estimate the amplitude of electric field necessary, if the energy required to twist an isolated water molecule from parallel to anti-parallel in the electric field, $2pE$, is equal to a typical thermal energy at $20\,°C$ of 4.05×10^{-21} J.

4.4   A univalent potassium ion of radius 0.133×10^{-9} m is immersed in water with a bulk relative dielectric constant of $\varepsilon_R = 80$. Assume that a water molecule is spherical with a radius of 0.14×10^{-9} m and has an electric dipole moment of 6.17×10^{-30} C m located at its centre. Calculate the electric field at the dipole moment due to the charge of 1.6×10^{-19} C on the ion when they are in contact, and hence calculate the energy required to rotate the dipole from parallel to anti-parallel to this field on the assumption that (a) the water surrounding the ion is a continuous fluid with $\varepsilon_R = 80$ and (b) that $\varepsilon_R = 1$. Discuss the plausibility of the two assumptions (a) and (b).

4.5*   Show that within the body of a homogeneous dielectric there can be no bound charge density where there is no free charge density.

Worked solution

Within the body of a homogeneous dielectric we have the relation $D = \varepsilon_R \varepsilon_0 E$. If we apply Gauss's law in its various forms to a small volume dV containing a free charge density ρ_F and a bound charge density ρ_B, we obtain

$$\iiint_V \rho_F dV = \iint_S D_N dS = \varepsilon_R \varepsilon_0 \iint_S E_N dS$$

$$= \varepsilon_R \varepsilon_0 \iiint_V \frac{\rho_F + \rho_B}{\varepsilon_0} dV \quad \text{or} \quad \rho_F = \varepsilon_R (\rho_F + \rho_B).$$

Thus when $\rho_F = 0$ we must also have $\rho_B = 0$ as required.

The solution is more compact if we use the microscopic differential forms of Gauss's law introduced in Problem 1.5:

$$\text{div}(\tilde{E}) = \frac{\rho_F + \rho_B}{\varepsilon_0} \quad \text{and} \quad \text{div}(\tilde{D}) = \rho_F.$$

so that we must have

$$\rho_F = \text{div}(\tilde{D}) = \text{div}(\varepsilon_R \varepsilon_0 \tilde{E}) = \varepsilon_R \varepsilon_0 \text{div}(\tilde{E}) = \varepsilon_R \varepsilon_0 \frac{\rho_F + \rho_B}{\varepsilon_0}.$$

which leads to the same conclusion. Note that on the surface of a dielectric there can exist bound charge when no free charge is present.

4.6   An electric field of amplitude $\tilde{E}_1$ in a material of relative dielectric constant ε_{R1} meets the plane interface with a second dielectric material with relative dielectric constant ε_{R2} perpendicular to the interface. Obtain an expression for the bound surface charge density σ at the interface using (a) Gauss's law and (b) the relation $\tilde{D} = \varepsilon_0 \tilde{E} + \tilde{P}$ together with equation [4.7]. Give a physical explanation why the surface charge density is positive if $\varepsilon_{R1} > \varepsilon_{R2}$.

4.7   A uniform electric field $\tilde{E}$ is present in a solid material of relative dielectric constant ε_R. There are two cavities in the material. One is a long needle-shaped cavity with its axis parallel to the field, and the other is a thin disc-shaped cavity with the plane disc perpendicular to the field. Using the boundary conditions for $\tilde{E}$ and $\tilde{D}$, obtain expressions for the electric field in the centre of both cavities. Where on the cavity walls would you expect to find bound surface charge due to the action of the field $\tilde{E}$? Show that the calculated electric field amplitude within the two cavities is the sum of the original field and the extra field due to the surface charge density.

5. The calculation of electric fields and voltages in the presence of dielectrics

In this chapter the techniques developed in Chapter 4 to deal with the presence of dielectric material, are applied to situations involving charges, conductors and dielectrics. In particular, the enhancement of the capacitance of capacitors filled with dielectrics is discussed. Also the electrostatic self-energy of a single charged conductor is derived together with the extra electrostatic energy stored in an assembly of interacting charged conductors. An expression for the energy per unit volume stored in an electric field is deduced. The reason why the plasma membrane of an animal cell is such an effective barrier to the passage of small ions across it is shown to be electrostatic in origin.

5.1 The dielectric constant of water

To give some physical insight into dielectric behaviour we will briefly discuss the origins of the dielectric constant of water which is of paramount importance in biology. The relative dielectric constant of water at room temperature is about 80, which is among the highest of any known fluid. As we shall see, this has a profound effect on electric fields generated in water and hence on the behaviour of ions in solution. From equation [4.15] we may deduce a relation between the dipole moment per unit volume $\tilde{P}$ induced in water and the inducing electric field $\tilde{E}$, so that

$$\tilde{P} = (\varepsilon_R - 1)\varepsilon_0\tilde{E}. \qquad [5.1]$$

Thus for an applied electric field of $1000\,\mathrm{V\,m^{-1}}$ the dipole moment per unit volume induced in water becomes $6.99 \times 10^{-7}\,\mathrm{C\,m^{-2}}$. As mentioned in Section 4.3, the reason for the large relative dielectric constant in water is that each water molecule has a relatively large electric dipole moment of $6.17 \times 10^{-30}\,\mathrm{C\,m}$ which may rotate in response to an applied electric field. There are about 3.35×10^{28} water molecules in a cubic metre (55.6 molar) so that to exhibit a dielectric constant of about 80 in an electric field of $1000\,\mathrm{V\,m^{-1}}$ requires an induced dipole moment per molecule in the direction of the applied field of about $2.09 \times 10^{-35}\,\mathrm{C\,m}$. This is only about 1 part in

290 000 of the permanent dipole moment of each molecule so that the alignment created by even a strong applied electric field of 1000 V m^{-1} is very small indeed. The same induced dipole moment per unit volume would be created if 1 molecule in each 290 000 molecules rotated into the direction of the applied electric field while the others remained randomly oriented with no net dipole moment. Of course the real time-averaged molecular alignment in the field is equally shared by all the molecules and is so small because of the disturbing effects of collisions with neighbouring water molecules. Despite this apparently small response, a fluid with a dielectric constant of 80 reduces the electric fields and voltages surrounding a charge distribution by a large factor of 80, as we shall see below.

5.2   The electric field and voltage of a charged conducting sphere within dielectric material

Let us consider the case of a small spherical conductor of radius a carrying a total uncompensated free charge of $+Q$ embedded in a large block of homogeneous dielectric with a relative dielectric constant of ε_R as sketched in Fig. 5.1. This system

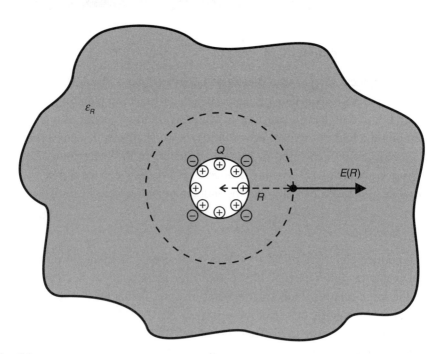

Fig. 5.1   A spherical conductor of radius a and carrying a total charge Q embedded in a large block of material with a relative dielectric constant ε_R. The electric field $E(R)$, a distance R from the centre of the conductor, such that $R > a$, is shown.

is clearly spherically symmetric about the centre of the spherical conductor. If we draw a spherical Gauss law surface centred on the centre of the conductor with radius R ($R > a$), we may calculate the flux of $\tilde{D}$ through the surface as $4\pi R^2 D(R)$. Applying Gauss's law as in equation [4.12], we obtain

$$4\pi R^2 D(R) = Q \quad \text{or} \quad D(R) = \frac{Q}{4\pi R^2}. \qquad [5.2]$$

Then, using the condition that $D = \varepsilon_R \varepsilon_0 E$ as in equation [4.15], we deduce that

$$E(R) = \frac{Q}{4\pi\varepsilon_R\varepsilon_0 R^2}. \qquad [5.3]$$

Thus, in the presence of the dielectric, the electric field has the same shape as in a vacuum but the amplitude of the field is reduced by a factor ε_R. We can understand this physically as being due to the negative surface polarization charge $-P_N$ induced on the inner surface of the dielectric adjacent to the charged sphere. We deduce that the total charge within the spherical Gauss law surface, free charge plus induced bound charge, has been reduced by a factor of ε_R which accounts for the reduction of the electric field. Using the same arguments as in Chapter 1, we can deduce that the voltage in the dielectric surrounding the conductor is also reduced by the same factor and becomes

$$V(R) = \frac{Q}{4\pi\varepsilon_R\varepsilon_0 R}. \qquad [5.4]$$

The expressions for $E(R)$ and $V(R)$ have the same R-dependence as in a vacuum so that we can deduce that there is no body charge $-\text{div}[\tilde{P}(x, y, z)]$ induced within the dielectric to create or terminate the electric field. We have assumed a uniform dielectric where the amplitude of P is proportional to the amplitude of E, so that to prove $\text{div}[\tilde{P}] = 0$ requires only that we prove $\text{div}[E] = 0$ at a general position (x, y, z) within the dielectric. The cosine of the angle between the x-axis and any vector $\tilde{R}$ from the centre of the charge to the point (x, y, z) is simply x/R, so that the component of the field at that point in the x-direction is given simply by $E_x = (K/R^2)(x/R)$ using equation [5.3], where K is the constant $Q/4\pi\varepsilon_R\varepsilon_0$. To calculate dE_x/dx we use the fact that $R^2 = x^2 + y^2 + z^2$, so that $(dR/dx) = x/R$, and calculate $d(x/R^3)/dx$ as follows:

$$\frac{d}{dx}\left(\frac{x}{R^3}\right) = \frac{1}{R^3} - \frac{3x}{R^4}\left(\frac{dR}{dx}\right) = \frac{1}{R^3} - \frac{3x^2}{R^5} = \frac{R^2 - 3x^2}{R^5}.$$

Thus we obtain

$$\frac{dE_x}{dx} = K\frac{(R^2 - 3x^2)}{R^5},$$

so that, dealing similarly with the other components of $\tilde{E}$, we obtain

$$\text{div}[\tilde{E}] = K\left(\frac{dE_x}{dx} + \frac{dE_y}{dy} + \frac{dE_z}{dz}\right) = \left(\frac{K}{R^5}\right)[3R^2 - 3(x^2 + y^2 + z^2)] = 0 \qquad [5.5]$$

as required. Note that in this case the electric field and hence P are varying with R, and the cancellation of the body charge density to zero depends upon the fact that the volume between R and $R + dR$ increases with R as R^2 so as to exactly compensate for the fact that P varies with R as R^{-2}, because $E(R)$ varies as R^{-2}.

We can show quite generally that $\text{div}[\tilde{E}(x, y, z)] = 0$ within a homogeneous dielectric material, at a location (x, y, z) where there is no free charge by using the worked solutions to Problems 1.5 and 4.5. The solution to Problem 4.5 shows that there is no bound charge within a homogeneous dielectric where there is no free charge, so that the total charge density there is $\rho_{\text{tot}}(x, y, z) = 0$. The solution of Problem 1.5 results in the microscopic version of Gauss's law which states that

$$\text{div}[\tilde{E}(x, y, z)] = \frac{\rho_{\text{tot}}(x, y, z)}{\varepsilon_0}.$$

Combining the two results proves the equation that $\text{div}(\tilde{E}) = 0$ in the circumstances we defined above. Note that the result deduced in Problem 4.5 is not true at the **surface** of dielectric material where there can be bound charge without free charge.

The results that we have deduced in this section show that within a single homogeneous dielectric all the expressions we obtained for the electric field and voltage in a vacuum are unchanged in general form but simply have their amplitudes reduced by a factor ε_R. This clearly also holds for the field and voltage around an electric dipole and every other distribution of charges.

5.3   Electric fields at the interface between two different dielectrics

We will now discuss two examples which include an interface between two dielectrics. We discussed above the case of a spherical conductor of radius a on which is deposited a total uncompensated free charge of Q when the conductor was buried in an extended dielectric of relative dielectric constant ε_R. A simple extension of this case is when the dielectric is finite in volume and forms a thick spherical shell about the conductor with an inner radius of a and an outer radius of b, as sketched in Fig. 5.2. For distances from the centre of the charge R such that $a < R < b$ we obtain the same values of D and E as shown in equations [5.1] and [5.2]. However, when we apply Gauss's law over a spherical surface of radius $R > b$, the value of D remains the same because the only free charge within the Gauss law surface remains Q on the conductor.

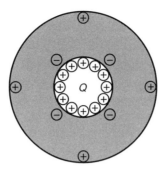

Fig. 5.2   A spherical conductor with a radius a and carrying a total charge Q is shown located within a thick spherical shell of material with a relative dielectric constant of ε_R. The dielectric material has an inner radius of a and an outer radius of b.

Thus the value of $E(R)$ for $R > b$ becomes $Q/4\pi\varepsilon_0 R^2$, because $\varepsilon_R = 1$ in this region. This is the result we would obtain in the absence of the thick dielectric shell and shows that the total charge, free and bound, within the Gauss law surface is simply Q. Thus we deduce that the positive surface polarization charge on the outer surface of the thick dielectric shell must equal the total negative surface polarization charge on its inner surface. We would expect this as the total body charge is zero ($\mathrm{div}[\tilde{P}] = 0$ within the dielectric) and thus the total surface charges must add to zero as the dielectric shell is isolated and thus must remain uncharged as a whole. There is a discontinuity in the amplitude but not in the direction of the radial electric field $E(R)$ at $R = b$ due to the presence of surface polarization bound charge, whereas the radial electric displacement D is continuous as there is no free charge on this surface.

Another example involving the interface between dielectrics is illustrated in Fig. 5.3. Here we show the plane interface between two homogeneous dielectrics with relative dielectric constants of ε_{R1} and ε_{R2}. We also show the electric fields in the two regions as $\tilde{E}_1$ and $\tilde{E}_2$, which make angles θ_1 and θ_2 with the normal to the interface. From the boundary conditions we derived we know that tangential $\tilde{E}$ is conserved so that

$$E_1 \sin(\theta_1) = E_2 \sin(\theta_2). \qquad [5.6]$$

We also know that the normal component of $\tilde{D}$ is conserved so that

$$\varepsilon_{R1} E_1 \cos(\theta_1) = \varepsilon_{R2} E_2 \cos(\theta_2). \qquad [5.7]$$

Dividing the two sides of equation [5.6] by the corresponding sides of equation [5.7], we obtain the result

$$\tan(\theta_1) = \frac{\varepsilon_{R1}}{\varepsilon_{R2}} \tan(\theta_2). \qquad [5.8]$$

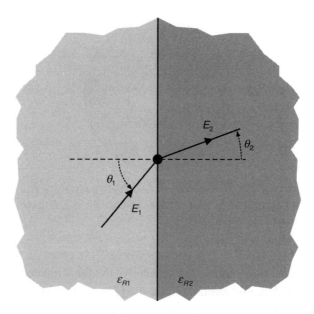

Fig. 5.3 The plane interface between material with relative dielectric constants ε_{R1} and ε_{R2} respectively. An electric field $\tilde{E}_1$, which meets the interface at an angle θ_1 to the interface normal, changes direction such that $\tilde{E}_2$ leaves the interface at an angle θ_2 to the interface normal.

Thus we show that the two angles, θ_1 and θ_2, are not equal and that **the electric field is diffracted at the plane interface between two dielectrics** with different relative dielectric constants.

5.4   Capacitors filled with dielectric material

Let us consider again the parallel plate capacitor with two conducting plates of large area A separated by a constant small distance d, but now with the space between the plates filled with a dielectric with a relative dielectric constant ε_R. We place an uncompensated free charge totalling $+Q$ on the upper plate and we connect the lower plate to Earth, as sketched in Fig. 5.4. Negative charge is attracted from the Earth to the lower plate. If we were to construct a cylindrical Gauss law surface with its flat ends buried in the two plates and its axis perpendicular to the plane of the plates, as in Fig. 2.5, we may deduce that the flux of $\tilde{D}$ through the Gauss surface is zero so that the total enclosed free charge is zero. This means that the positive free charge density on the lower surface of the upper plate is equal to the negative free charge density attracted to the upper surface of the lower plate. To prove this result we have used the fact that $\tilde{E}$ and thus $\tilde{D}$ in the dielectric are perpendicular to the surfaces of the plates. The argument for this supposition is to be found in Section 2.4. If we apply Gauss's law in $\tilde{E}$ to the

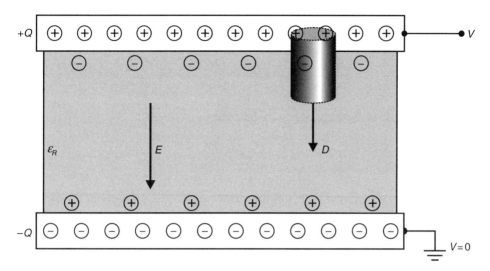

Fig. 5.4 A parallel plate capacitor filled with material of relative dielectric constant ε_R. The total charges on the inner surfaces of the upper and lower plane conducting plates are $+Q$ and $-Q$ and the voltage difference between the plates is V. The electric field within the dielectric material is $\tilde{E}$, which is directed normal to the plates. A cylindrical Gauss law surface buried in the upper plate is shown with its axis normal to the conducting plates.

same Gauss law cylinder we would deduce that the total charge, free and induced, within the cylinder is zero. Thus the negative bound surface charge density on the upper face of the dielectric must be equal in magnitude to the positive bound surface charge density on the lower face of the dielectric. These bound surface charges are due to the uncompensated ends of the electric dipole moments induced in the dielectric material by the downward-pointing electric field, as described in Section 4.3.

If we construct a cylindrical Gauss law surface as shown in Fig. 5.4, with its upper flat face buried in the upper plate and its lower flat face within the dielectric occupying the gap between the two conductors, we may deduce a relationship between D within the dielectric and the free charge density Q/A on the lower face of the upper conducting plate using Gauss's law:

$$D = \frac{Q}{A}.$$ [5.9]

The value of D deduced from equation [5.9] is the same no matter how far the Gauss law cylinder extends into the dielectric slab. Thus the amplitude of both D and the electric field $\tilde{E}$ are constant everywhere within the slab. The amplitude of the electric field is given by

$$E = \frac{D}{\varepsilon_R \varepsilon_0} = \frac{Q}{\varepsilon_R \varepsilon_0 A}.$$

Because the electric field is constant in amplitude the voltage between the plates is simply given by

$$V = Ed = \frac{Qd}{\varepsilon_R \varepsilon_0 A}.$$

The capacitance C of the capacitor now becomes

$$C = \frac{Q}{V} = \frac{\varepsilon_R \varepsilon_0 A}{d}, \qquad [5.10]$$

which is a factor of ε_R larger than was the case (equation [2.6]) without the dielectric. The physical reason is that the surface polarization charges on the outer plane faces of the dielectric slab partially cancel the free charges on the inner surfaces of the plates so that for a given voltage difference between the plates more free charge must be placed on the plates, thus increasing the capacitance. By an identical argument, the capacitance of the spherical capacitor discussed in Section 2.3 is also increased by a factor ε_R when the annular space between the spherical conductors is filled with a dielectric with a relative dielectric constant of ε_R. Increasing the capacitance of a capacitor as a charge storage device is a major practical use of material with a high relative dielectric constant.

5.5   The electrostatic self-energy of an isolated conductor

We have already discussed the energy of interaction between two electrical systems such as between two charges or between a charge and an electric field. Here we will calculate the **electrostatic self-energy** that any single isolated conductor possesses when it carries a charge. It is called a self-energy because it is due to the interaction of the charge on the conductor with its own electric field that is present because of that charge. We will apply this calculation below to estimate the self-energy of an ion in aqueous solution, and we discuss the importance to biology of this result in Section 5.8.

We first calculate the energy required to charge a conductor of any shape to a final voltage V_0 when it carries an uncompensated charge Q_0. As we discussed above and in Chapter 2, a conductor is an equipotential body and its voltage V is related to the uncompensated charge Q placed on it by a constant, which depends upon its shape and dimensions and is known as its capacitance C. The voltage on the conductor and the charge it carries are linearly connected such that $Q = CV$ as in equation [5.10] or [2.6]. We now consider the charging process starting with the conductor uncharged. At some intermediate time during the charging process let the voltage of the conductor

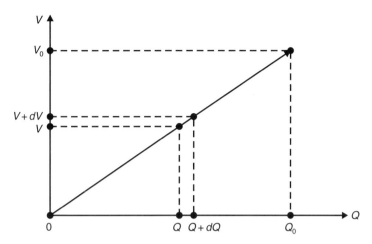

Fig. 5.5   A graph of the voltage V on an isolated conductor shown as a function of the charge Q brought up to charge the conductor. Because the capacitance of the capacitor remains constant, the voltage is proportional to the charge brought up. The final voltage and charge are V_0 and Q_0.

be $V = fV_0$ so that the charge it carries must be $Q = fQ_0$, where f is a fractional number between 0 and 1. The situation is illustrated by Fig. 5.5, which shows how the voltage grows linearly with the charge. If we now bring up the next small increment of charge $dQ = Q_0df$, the incremental work done by the external agency moving the charge dW is equal to $+VdQ$, as the charge dQ is moved from far away where the voltage is zero to the surface of the conductor where the voltage at this stage is V. Substituting for V and dQ, we find that $dW = dQ \cdot V = Q_0 \cdot df \cdot V_0 \cdot f$ and the total work done W in charging the conductor to its final voltage V_0 may be calculated by adding all the increments dW by integration, from the start of the charging process when $f = 0$ to the end when $f = 1$, so that we have

$$W = \int_{f=0}^{f=1} dW = \int_{f=0}^{f=1} Q_0V_0fdf = Q_0V_0\int_{f=0}^{f=1} fdf = \frac{1}{2}Q_0V_0. \qquad [5.11]$$

The reason for the factor 1/2 is clearly seen in Fig. 5.5. The incremental work done dW is represented by the small rectangle of area VdQ and the total work done is the sum of all such incremental areas which becomes the area of a triangle of base length Q_0 and height V_0 which is $\frac{1}{2}Q_0V_0$. The total work W done in charging the conductor is the **electrostatic self-energy** of the charged conductor $U_S = \frac{1}{2}Q_0V_0$.

The argument above clearly holds for a conducting body of any shape or size. As an example important in biology, we will consider the particular case of the self-energy of an ion in aqueous solution. For this discussion we will assume that an ion is a small conducting sphere of radius a that carries a total uncompensated charge of Q_0. When the spherical conductor is immersed in a fluid of relative dielectric constant ε_R, the voltage around it is reduced by a factor ε_R as in equation [5.4], so that at its surface when $R = a$ the voltage is given by

$$V_0 = \frac{Q_0}{4\pi\varepsilon_R\varepsilon_0 a}. \tag{5.12}$$

The electrostatic **self-energy** of the ion U_S, which measures the energy stored in the ion because of the interaction between its charge and the electric field generated by its own charge, is given by

$$U_S = W = \frac{Q_0 V_0}{2} = \frac{Q_0^2}{8\pi\varepsilon_R\varepsilon_0 a}. \tag{5.13}$$

We can express the self-energy of a conductor in several ways using the relationship $Q_0 = CV_0$ between the charge Q_0, the voltage V_0 and the capacitance C:

$$U_S = \frac{Q_0 V_0}{2} = \frac{CV_0^2}{2} = \frac{Q_0^2}{2C}.$$

The importance for biology of these expressions for the self-energy of an ion in a dielectric environment, which will be discussed further in Section 5.8, is that for ions with small radius a, the self-energy U_S can be very large.

Omitting the factor of one-half in the expression for the self-energy of an ion is probably the most common electrostatic error made in the biological literature. It arises through the false argument that the voltage is V_0 and the charge brought up is Q_0 so that the work done is $Q_0 V_0$. The error in this argument is the assumption that the voltage remains constant as the charge is brought up. As can be seen in Fig. 5.5, the work done to bring up the first increment of charge is zero because the voltage is then zero. The work then grows for subsequent increments of charge as the body becomes charged and its voltage rises. However, only the last increment has to perform the work required to overcome the final voltage of V_0. A self-energy can always be recognized when the force against which work is done, when bringing up the increments of charge, is the electric field created by the presence of the accumulating charge itself. Thus to charge a previously uncharged conductor to a final voltage of V_0, when it carries a charge of Q_0, requires an amount of work $U_S = Q_0 V_0/2$ as in equation [5.13]. However, to bring a charge Q_0 up to a conductor which is maintained at a voltage V_0 by an external agency, such as a battery, requires work $U = Q_0 V_0$ because all the charge brought up is raised to a fixed voltage of V_0.

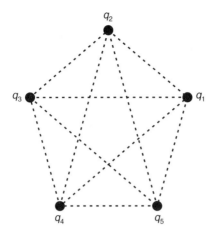

Fig. 5.6   Five charges located in a plane showing as broken lines the interactions between the charges that store electrical energy.

5.6   The extra electrostatic energy of an assembly of charges

Here we are concerned with a number of different charges, which we may think of as small charged conductors, that are brought together such that they interact. Each charge when isolated from all the others has an electrostatic self-energy, as we have discussed above. We will now calculate the **extra electrostatic energy** ΔU that arises because of the interactions between the charges when they are brought closer together. For these purposes $\Delta U = 0$ represents the energy of the array when the charges are widely separated. The important physical principle is that contributions to ΔU arise from the electrostatic interactions between each pair of charges and we must take care to count each of these interactions only once. We consider the extra energy ΔU stored in an assembly of i individual charges q_i such that the voltage at the ith charge due to all the **other charges** is V_i. Such a situation is shown in Fig. 5.6 for five charges, where the 10 energy-storing interactions are shown as broken lines. It may at first be thought that we need only sum the product $q_i V_i$ over all the charges to obtain the energy of the array, but to do so would count each interaction between the charges twice. The correct value of the energy of the array ΔU is given by

$$\Delta U = \frac{1}{2} \sum_i q_i V_i. \tag{5.14}$$

The voltage V_i is the sum over all values of j of the separate voltages V_{ji} generated at the ith charge by the jth charge. When we evaluate the term $q_i V_i$ in equation [5.14], the term $q_i V_{ji}$ represents the energy of interaction between the ith and jth charges.

However, when performing the summation in equation [5.14] we must also evaluate the term $q_j V_j$, where V_j is the sum of separate voltages V_{ij} generated at the jth charge by the ith charge. The term $q_j V_j$ will contain the term $q_j V_{ij}$ which again represents the energy of the interaction between the ith and jth charges. Thus without the factor of one-half in equation [5.14] we would count the interaction energy between each two charges twice.

5.7   The energy stored in an electric field

An important application of the concept of a self-energy is that of the capacitor, and we will consider as an example the parallel plate capacitor as sketched in Fig. 5.4 with a plate area A and a plate separation d. The stored energy arises because work must be done in delivering charge to the top plate up a voltage gradient from zero volts, very far away, to a voltage V on the upper plate, just as we discussed in Section 5.5 for an isolated spherical conductor. We recognize a self-energy because the delivered charge has an energy due to the presence of the voltage the charge itself creates. If the total final uncompensated charge on the upper plate is $+Q_0$ and the final voltage between the plates is V_0 then the energy stored in the capacitor is given by U_C, which, using equation [5.13], becomes

$$U_C = \frac{1}{2} Q_0 V_0. \qquad [5.15]$$

Thus far we have expressed the extra electrostatic energy of an electrical array as residing in the individual charged bodies present, interacting with the local prevailing voltages whether self-generated or from external sources. This is known as the **source model of electrical interaction energy**.

An equally valid approach is to say that although electric fields require no energy to maintain they do require energy to create, and that the energy of an electrical array resides in the electrical fields it creates between the charged bodies. This is known as the **field model of electrical interaction energy**. Both viewpoints are possible and in fact are exactly equivalent. We can use either approach to calculate stored electrical energy, but we must be careful to use one approach or the other and not both or we will get an answer twice as big as it should be. We will use the example of the energy stored in the parallel plate capacitor to deduce the energy stored in an electric field per unit volume.

Equation [5.15] gives the energy stored in the parallel plate capacitor using the source model. We know from Section 5.4 that $D = Q_0/A$, that $E = D/\varepsilon_0\varepsilon_R$ and that $V_0 = Ed$ where E and D are constant within the capacitor. Thus we may write the stored energy in terms of the fields E and D as

$$U_C = \frac{Q_0 V_0}{2} = \frac{DAEd}{2} = \frac{DE}{2} Ad$$

$$= \frac{DE}{2} \quad \text{(Volume of the capacitor).} \quad [5.16]$$

The electric field E has associated with it an electrical energy per unit volume or **energy density** within a capacitor of ρ_E, where

$$\rho_E = \frac{DE}{2} = \frac{\varepsilon_R \varepsilon_0 E^2}{2}. \quad [5.17]$$

The energy density we have calculated for the particular case of a capacitor holds generally for all electrical fields, however caused. Thus, in general, **the energy stored per unit volume at a position where the electric field is $\tilde{E}$ and the electric displacement is $\tilde{D}$ is $\tilde{D} \cdot \tilde{E}/2$**. It is a matter of choice in a particular situation whether to use the field model or the source model when calculating stored electrical energy.

5.8   The cell plasma membrane as a barrier to ions

The fact that the plasma membrane of animal cells is such an effective barrier to the passage of small ions is due to the large difference in the electrostatic self-energy of the ion in aqueous solution and within the membrane. Using equation [5.13], we may calculate the rise in self-energy, ΔU_S, of an ion of radius a and charge q when it moves from an aqueous solution with relative dielectric constant ε_{RW} to the membrane with its relative dielectric constant of ε_{RM} as

$$\Delta U_S = \frac{q^2}{8\pi\varepsilon_0 a} \left(\frac{1}{\varepsilon_{RM}} - \frac{1}{\varepsilon_{RW}} \right). \quad [5.18]$$

The relative dielectric constant for water is given by $\varepsilon_{RW} = 80$ and is among the highest known for any fluid because of the powerful permanent electric dipole moments of the water molecules that may rotate under the influence of the applied electric field. Within the membrane, where there are no rotating electric dipoles in an applied electric field, ε_{RM} is only about 2 because of slight electron cloud distortion in an electric field. If we take as an example the univalent potassium ion which carries a charge 1.6×10^{-19} C and has a bare radius a of about 1.38×10^{-10} m, the value of ΔU_S becomes 4.06×10^{-19} J, which is about 100 times larger than a typical thermal energy at 20 °C of $kT = 4.04 \times 10^{-21}$ J.

Our calculation is unrealistic in several respects. It assumes that the water is a continuous fluid whereas it is granular, with the molecular diameter of water comparable to that of the ion, which limits the closest approach of water to the ion. We have also assumed that the dielectric constant of water is 80 even very close to the ion, whereas it is likely that the nearest-neighbour water molecules are held immobile by the large electric field of the ion, which limits their response to the field and thus limits their effective relative dielectric constant. However, the experimentally derived value for the hydration energy for the potassium ion, which is the energy required to remove it from water to a vacuum, is about 5.33×10^{-19} J, which would correspond to a value about 2.63×10^{-19} J in the move from water to a region with relative dielectric constant of 2. Thus our simple calculation has exaggerated the change in energy of the ion by a factor of about 1.5 but it does give the right order of magnitude and rightly points to the electrostatic origin of these very large energies.

Thus the reason why the plasma membrane is such an effective barrier to small ions is directly electrostatic in origin and is not steric. Small ions could easily fit in the cavities created by the thermally activated motion of the hydrocarbon tails of the lipids that compose the plasma membrane, but they are excluded by their very high electrostatic self-energies in this environment. The probability of a spontaneous transfer from water to the lipid at 20 °C may be estimated (see Appendix 2) by the Boltzmann factor of $\exp(-\Delta U_S/kT)$, where ΔU_S is the energy rise on transferring the ion from water to the membrane. Using the experimentally derived value of ΔU_S mentioned above, this probability for a thermally activated transfer from water to the membrane becomes about $\exp(-65)$, which is negligibly small. Ion channels which convey ions across the membrane in a controlled manner must somehow lower the self-energy of the ions during transit if they are to be effective. We will discuss some of the properties of ion channels in Chapter 13.

Topics covered in Chapter 5

1.   The relative dielectric constant of water is about 80, which is among the highest of any fluid. The reason for this high value is the large number of water molecules in unit volume and the large permanent electric dipole moment of each water molecule which becomes partially aligned in an applied electric field.

2.   The electric field and the voltage surrounding a single charge embedded in a homogeneous dielectric material have the same dependence upon the radial distance from the charge as in a vacuum, but their amplitudes are reduced by division by the relative dielectric constant. In the case of water this could mean a reduction by a factor of about 80.

3.   Even in a fluid with a large relative dielectric constant such as water, the induced electric dipole moment per unit volume is very small in comparison with the saturated electric dipole moment that would be obtained if all the electric dipoles were perfectly aligned.

4.   If the space between the conducting plates of a capacitor, such as the parallel plate capacitor, is filled with material with a relative dielectric constant of ε_R, the capacitance of the capacitor is increased by a factor of ε_R.

5.   At the plane interface between two materials with different relative dielectric constants, the direction of an electric field that makes an angle with the surface normal, is diffracted so that it abruptly changes direction.

6.   An isolated charged conductor has an electrostatic self-energy due to the interaction of the charge on its surface and the electric field that the charge creates. The size of this self-energy is given by the uncompensated charge on the conductor multiplied by its voltage and divided by 2. The factor of 2 arises because the charges interact with the electric field which these same charges create.

7.   The electrostatic self-energy of small ions can be very large in a vacuum because the electric field at its surface is large. This self-energy is much reduced when the ion enters water because the high relative dielectric constant of the water reduces markedly the electric field at the surface of the ion.

8.   The reason why the plasma membrane of an animal cell is such an effective barrier to the passage of small ions is that the electrostatic self-energy of the ions is so much higher in the membrane than it is in water. The hydrocarbon tails of the lipids that compose the membrane have a low electrical polarizability and hence a low relative dielectric constant.

9.   The electrostatic interaction energy of an assembly of charged conductors can be thought to reside in the interaction of the charges on the surfaces of the conductors with the electric fields created by all the other charges. This approach is called the source model of the interaction energy. An exactly equivalent view is that the energy is stored in the electric fields that exist in the spaces between the charged conductors. This is called the field model of the interaction energy. In such calculations care must be taken to use one model or the other but not both, which would yield twice the correct interaction energy.

Important equations of Chapter 5

1.   The electric field $E(R)$ and the voltage $V(R)$ at a distance R from the centre of a single charge Q immersed in a homogeneous material of relative dielectric constant ε_R are reduced by a factor of ε_R over those in a vacuum and are given by

$$E(R) = \frac{Q}{4\pi\varepsilon_R\varepsilon_0 R^2} \quad \text{and} \quad V(R) = \frac{Q}{4\pi\varepsilon_R\varepsilon_0 R}.$$

2.   The capacitance of the parallel plate or the spherical capacitor is increased by a factor ε_R by filling the space between the two conductors with material of dielectric constant ε_R.

3.   The electrostatic self-energy of a conductor charged to a voltage V which carries an uncompensated charge of Q is given by

$$U_S = \frac{QV}{2}.$$

4.   The extra electrostatic energy ΔU stored in an array of electrically interacting charges q_i is given by

$$\Delta U = \frac{1}{2} \sum_i q_i V_i,$$

where V_i is the voltage at the ith charge created by all the other charges.

5.   The energy density (energy per unit volume) stored at a position where the electric field is $\tilde{E}$ and the electric displacement is $\tilde{D}$ is given by

$$\rho_E = \frac{\tilde{D}\tilde{E}}{2}.$$

Within a homogeneous dielectric material with a relative dielectric constant of ε_R the energy density is given by

$$\rho_E = \frac{\varepsilon_R \varepsilon_0 E^2}{2}.$$

6.   The barrier that prevents the flow of small ions across the lipid plasma membrane of an animal cell is due to the rise in electrostatic self-energy that the ion would experience in a move from water with a high relative dielectric constant to the lipid with a low relative dielectric constant. A very approximate expression for the rise in self-energy of an ion of radius a and charge q which moves from a region with a relative dielectric constant ε_{R1} to a region with a relative dielectric constant ε_{R2} is

$$\Delta U_S = \frac{q^2}{8\pi\varepsilon_0 a} \left(\frac{1}{\varepsilon_{R2}} - \frac{1}{\varepsilon_{R1}} \right).$$

Problems

5.1   A plane parallel-sided sheet of soda glass, with a relative dielectric constant $\varepsilon_R = 7$, is oriented perpendicular to the direction of a uniform applied electric field of $E_0 = 10^4$ V m^{-1}. What is the magnitude of the electric field inside the sheet and the induced electric dipole moment per unit volume P induced within the glass? What is the surface charge per unit area induced on the two plane faces of the sheet? If the sheet has a thickness of 1 mm, what is the electric dipole moment per unit area induced in the sheet?

5.2   Charges Q_1, Q_2, Q_1, Q_2 are placed in a vacuum at the corners of a plane square of side a such that equal charges occupy opposite corners of the square. Calculate the total energy stored in the electrical interactions between the charges.

5.3   An electric field is directed into the plane face of a block of glass with relative dielectric constant $\varepsilon_R = 7$ such that it makes an angle θ_1 with the normal to the surface. Within the glass the field makes an angle θ_2 with the surface normal. Calculate the values of θ_2, when θ_1 takes values of $0°, 20°, 40°, 60°$ and $80°$. Hence show that no matter what the magnitude of θ_1, the electric field in the second material makes an angle θ_2 less than $90°$ with the interface normal.

5.4   A spherical cavity of radius a is cut in a large block of solid material with a relative dielectric constant ε_{R2}. A charge Q is placed at the centre of the cavity which is then filled with a liquid of relative dielectric constant ε_{R1}. Use Gauss's law to obtain an expression for Q_B, the total bound charge induced at the cavity surface. Show that Q_B must have the same sign as Q if $\varepsilon_{R1} > \varepsilon_{R2}$. Give a physical explanation for this fact. Show that the rise in electrostatic energy due to the interaction between Q and the induced bound charge is given by $\Delta U = (Q_B/2)(Q/4\pi\varepsilon_0\varepsilon_{R2}a)$. A calculation similar to this is used in Chapter 13 to calculate the excess electrostatic energy of an ion in a narrow water-filled cylindrical pore that spans a biological membrane.

5.5*   A parallel plate capacitor has conducting plates of large area A separated by a small distance d. The space between the plates is filled with material of relative dielectric constant ε_R to a height t ($t < d$) above the bottom plate. The lower plate is connected to Earth and the upper plate bears a total charge $+Q$. Find the voltage on the top plate and the capacitance of the capacitor.

Worked solution

The capacitor is sketched in Fig. 5.7. Using the symmetry of the problem we will assume that the fields $\tilde{E}$ and $\tilde{D}$ within the capacitor have a direction normal to the plates, as we did in Section 5.4. Applying Gauss's law to the cylindrical surface, marked in the figure with plane ends of area dS and axis perpendicular to the plates, we find

$$DdS = \left(\frac{Q}{A}\right)dS \quad \text{or} \quad D = \frac{Q}{A},$$

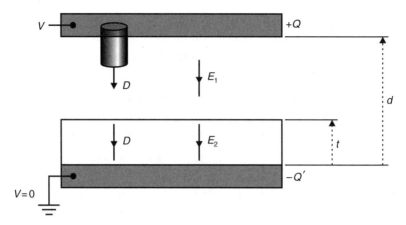

Fig. 5.7 A parallel plate capacitor with a plate separation d which has its lower plate covered by a parallel-sided slab of dielectric of thickness t and relative dielectric constant ε_R. The voltage difference between the plates is V and the surface charge on the inner face of the upper plate is $+Q$. In the air space the electric field inside the capacitor is E_1 and within the dielectric slab it is E_2. A cylindrical Gauss law surface with its axis normal to the plates is shown penetrating through the lower surface of the upper plate.

as in equation [5.9]. Normal $\tilde{D}$ is conserved so that D has the same value Q/A within the dielectric. Thus the values of the electric field above the dielectric and in the dielectric become

$$E_1 = \frac{D}{\varepsilon_0} = \frac{Q}{A\varepsilon_0} \quad \text{and} \quad E_2 = \frac{D}{\varepsilon_R\varepsilon_0} = \frac{Q}{A\varepsilon_R\varepsilon_0}.$$

Because these fields are constant in these two regions, the voltage across the capacitor V becomes

$$V = E_1(d - t) + E_2t = \frac{Q}{A\varepsilon_0}\left[(d - t) + \frac{t}{\varepsilon_R}\right].$$

The required capacitance C then becomes

$$C = \frac{Q}{V} = \frac{A\varepsilon_0}{(d - t) + \dfrac{t}{\varepsilon_R}}.$$

Note that the total positive charge density on the lower surface of the top plate is no longer equal to the total charge density at the upper surface of the bottom plate because there is positive bound charge on the lower plane surface of the material slab. Using a long cylinder with both its ends buried in the plates as in Fig. 2.5, we can prove, using Gauss's law in $\tilde{D}$, that the total free charge in that cylinder is zero. Thus the positive **free** surface charge density on the lower surface of the upper plate has a magnitude equal to the negative **free** surface charge density on the upper surface of the lower plate. Applying Gauss's law in $\tilde{E}$ to the same cylinder, we can show that the negative **bound** surface charge density on the upper surface of the dielectric slab has a magnitude equal to the positive **bound** surface charge density on the lower surface of the dielectric slab.

5.6   A voltage difference V is maintained between the plates of a parallel plate capacitor by connection to a battery. A slab of dielectric material is inserted between the plates. During the insertion does charge flow from the capacitor to the battery or vice versa?

5.7   Two uncharged capacitors of capacity C_1 and C_2 are connected in series between two terminals. Show that the circuit between the terminals acts as a capacitor of capacity C such that

$$\frac{1}{C} = \frac{1}{C_1} + \frac{1}{C_2}.$$

5.8   Using the formula derived in the previous problem, show that the capacitor discussed in Problem 5.5 is equivalent to a parallel plate capacitor with plate area A and plate separation t filled with material with relative dielectric constant ε_R connected in series with a second parallel plate capacitor with plate area A and plate separation $d - t$ which has no material between its plates. Give a physical explanation of this fact.

5.9   A parallel plate capacitor has a plate area A, a plate separation x and thus a capacitance of $C = A\varepsilon_0/x$. The fixed lower plate is connected to the Earth and the electrically isolated upper plate carries a fixed total charge $+Q$. A force F is applied to the top plate to counteract the attraction between the opposite charges on the two plates and maintain the plate separation as x. The electrical energy stored in the capacitor is given by $U = \frac{1}{2}Q^2/C$. If the plate separation increases by a small distance dx to become $x + dx$, the work done by the force F is $F \cdot dx$. When x increases by dx the stored energy increases by

$$dU = \left(\frac{Q^2}{2}\right) d\left(\frac{1}{C}\right) = \left(\frac{Q^2}{2}\right) d\left(\frac{x}{A\varepsilon_0}\right) = \left(\frac{Q^2}{2}\right)\left(\frac{1}{A\varepsilon_0}\right) dx.$$

By equating the work done by the force Fdx to the increase dU in the energy stored in the capacitor, find an expression for force F and thus find the force of attraction between the plates. Transform your expression for the force acting between the plates so that it is given in terms of the charge Q on the upper plate and the electric field E between the plates. Explain the factor of 1/2 that appears in your answer.

5.10   Make an estimate of the capacitance of a 1 mm length of the outer membrane of a squid giant axon which has a diameter of 0.5 mm. The membrane is non-conducting and has a thickness of 5×10^{-9} m and a relative dielectric constant of 2. Estimate the number of univalent cations that must enter the 1 mm length to raise the internal voltage by 50 mV.

6. Static magnetic fields

This chapter introduces magnetism. Magnetic fields are generated when currents flow in conductors; in particular, the magnetic field of a small loop of current-carrying conductor is equivalent to that of a magnetic dipole. There are no magnetic unipoles corresponding to electric charges. The two fundamental laws which govern the properties of magnetic fields, Gauss's law and Ampère's law, are derived. A magnetic potential is introduced and the differences between the electrostatic potential and magnetostatic potential are discussed.

6.1 The magnetic field and Gauss's law

We believe that magnetic effects play a very much smaller role in biological function than electric effects. The mobile ions on which life depends are all acted upon directly by electric fields but there is little in a living cell that responds directly to magnetic fields. A notable exception is that many living systems are known to use the weak magnetic field of the Earth as an aid to navigation, and this is treated as a specialist topic in Chapter 14. Despite this apparent lack of direct applicability to biological function, we will develop the theory of magnetism in the next few chapters both because time-varying magnetic fields can generate electric fields within a cell, as described in Chapter 9, and because the distinction between electric and magnetic effects is not as clear-cut as may at first be thought, as discussed in Chapter 10.

In the development of the theory of electricity we started with the experimental observation of the Coulomb force between two charges and action at a distance. We then defined the electric field created by a single charge and also defined the force exerted by this field on a second charge in a manner that was consistent with the experimentally observed Coulomb force. In magnetism the equivalent experimental observation is that there is a **force between two current-carrying conductors** even in a vacuum. In this case the magnetic field $\tilde{B}$ is defined by the statement that a short piece of a conductor of length dL metres which carries a current of I amperes will experience a force of dF newtons when placed in a magnetic field of amplitude B, where

$$dF = I \cdot B \cdot dL \cdot \sin(\theta). \qquad [6.1]$$

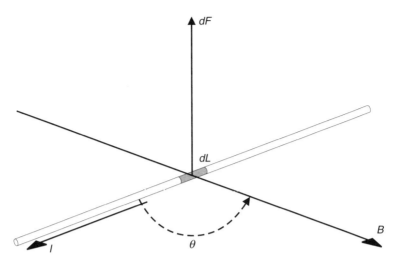

Fig. 6.1   A conducting wire carrying a current $\tilde{I}$ that makes an angle θ to the direction of the local magnetic field $\tilde{B}$. The force caused by the interaction between the current flowing in a short section dL of the wire and the magnetic field $\tilde{B}$ is shown as $d\tilde{F}$. The elemental force $d\tilde{F}$ is perpendicular to the plane containing the current $\tilde{I}$ and the magnetic field $\tilde{B}$.

The force is in a direction perpendicular to the plane that contains both $\tilde{B}$ and $\tilde{I}$ and points in the direction of advance of a right-handed thread rotated from the direction of $\tilde{I}$ to the direction of $\tilde{B}$. The angle θ is the angle between the directions of $\tilde{I}$ and $\tilde{B}$. The units in which B is expressed are then newtons per ampere per metre. The force dF acting upon a short length dL of a conductor carrying a current I in a magnetic field B is sketched in Fig. 6.1.

The result is most compactly expressed in terms of a **vector product** (see Appendix 1). The vector product, written as $[\tilde{a} \times \tilde{b}]$, between two vectors $\tilde{a}$ and $\tilde{b}$ is itself a vector with an amplitude of $ab\sin(\theta)$, where θ is the angle between the directions of $\tilde{a}$ and $\tilde{b}$. It acts in a direction perpendicular to both $\tilde{a}$ and $\tilde{b}$ and points in the direction of advance of a right-handed thread which is rotated from the direction of $\tilde{a}$ to the direction of $\tilde{b}$. Equation [6.1] may be written as

$$d\tilde{F} = [\tilde{I} \times \tilde{B}]dL, \qquad\qquad [6.2]$$

where the square bracket contains the vector product of $\tilde{I}$ and $\tilde{B}$.

If we were to follow exactly the course taken in our discussion of electricity we would define the magnetic field generated by a current-carrying conductor which, together with equation [6.1], would yield the force between two current-carrying conductors observed experimentally. However, more insight into magnetism is obtained if we combine equation [6.1] or [6.2] with the experimental observation that a small closed loop of current-carrying conductor produces a magnetic field with the same geometric shape as that of an electrical dipole. In Fig. 6.2a is shown a plane current loop with a small area dA and carrying a current I. Such a loop is defined as having a **magnetic dipole moment** with a magnitude given by

$$m = IdA. \hspace{4cm} [6.3]$$

The magnetic dipole axis is perpendicular to the plane of the small coil and the dipole points in the direction of advance of a right-handed screw, turned in the direction of the current in the loop. As can be seen from equation [6.3], the units in which the magnetic dipole moment is expressed are A m^2. Just as for an electric dipole shown in Fig. 4.1a, the magnetic field $\tilde{B}$ is in the form of closed loops which may in this case be considered to thread through the current loop.

If we use the same (R, θ) polar coordinates that we used in Fig. 4.1b for the electric dipole, we may define the components of the magnetic field at the field point $P(R, \theta)$ as $B_R(R, \theta)$ and $B_\theta(R, \theta)$, as sketched in Fig. 6.2b. These field components are given by the expressions

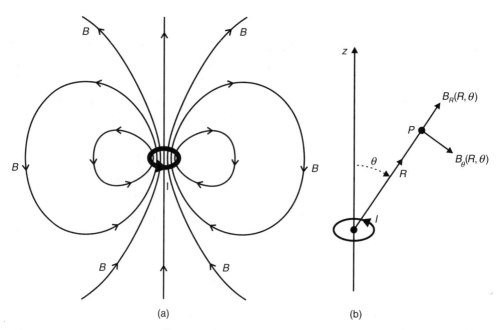

(a)   (b)

Fig. 6.2   (a) The magnetic field $\tilde{B}$ created by a small loop of conductor carrying a current I. The current-carrying loop is equivalent to a magnetic dipole of strength $\tilde{m}$ pointing in a direction normal to the loop. (b) The two-dimensional polar coordinates suitable for describing the magnetic field generated by a small current-carrying loop. The field point $P(R, \theta)$ is at a distance R from the loop and θ is the angle between the directions of $\tilde{R}$ and $\tilde{z}$, the normal to the loop. The diagram is in a plane that contains the field point $P(R, \theta)$ and the normal to the loop. The two components of the magnetic field $\tilde{B}$ at the field point $P(R, \theta)$ are shown as $B_R(R, \theta)$ and $B_\theta(R, \theta)$.

$$B_R(R, \theta) = \frac{2\mu_0 m \cos(\theta)}{4\pi R^3},$$

[6.4]

$$B_\theta(R, \theta) = \frac{\mu_0 m \sin(\theta)}{4\pi R^3}.$$

It can be seen by a comparison of equations [6.4] and equations [4.2] and [4.3], that these expressions for the magnetic field components have exactly the same R- and θ-dependence as the equivalent electric field components of the electric dipole. The constant μ_0 which is introduced is called the **permeability of free space** and has a value of $4\pi \times 10^{-7} \, \mathrm{N\,A^{-2}}$. It is the magnetic equivalent of the electrical constant ε_0. Note, however, that the magnetic constant μ_0 occurs in the numerator of the expressions in equation [6.4], whereas the electric constant ε_0 occurs in the denominator of the equivalent electrical expressions in equations [4.2] and [4.3]. As in the case of the electric dipole, the magnetic field at the point $P(R, \theta)$ lies entirely in the plane which contains the direction of the magnetic dipole and the field point $P(R, \theta)$. The full three-dimensional picture of the magnetic field is obtained by rotating a planar picture such as that in Fig. 6.2a around the axis of the dipole.

To date there is no experimental evidence for the existence of a **unit magnetic pole** that could take the place of a single electric charge. We believe that **all magnetic fields are created by current-carrying conductors, current loops or equivalently by magnetic dipoles** as described by equations [6.3] and [6.4]. We have seen above that the magnetic field generated by current loops or magnetic dipoles is in the form of closed loops, and we shall deduce below that the magnetic field created by any current-carrying conductor also forms closed loops. From this fact we may immediately deduce that the total outward flux of the magnetic field $\tilde{B}$ through any closed surface is zero. This is the **magnetic form of Gauss's law which states that the total flux of a magnetic field out of a closed surface of any shape is zero**. Just as in equation [1.4], we may write the flux as a surface integral where the surface is divided into a large number of small surface areas dS that exactly compose the total surface and where B_N is the component of the magnetic field normal to the surface area element dS:

$$\iint_S B_N dS = 0.$$

[6.5]

This law may also be considered to be the direct consequence of the absence of single magnetic poles.

6.2   The interaction of a magnetic dipole and a magnetic field

Figure 6.3 shows a small square current loop *abcda* centred on the origin and lying in the (x, y) plane. The sides *ba* and *cd* have lengths dx and the sides *cb* and *da* have lengths dy. The dipole moment $\tilde{m}$ points along the z-axis and has a value $I dx dy$, defined by equation [6.3], where I is the current circulating around the loop *abcda*. The magnetic field $\tilde{B}$ is chosen to lie in the (y, z) plane such that it makes an angle θ with the z-axis or equivalently to the direction of the magnetic dipole moment $\tilde{m}$.

Because the current flow is in opposite directions in *bc* and in *da*, we deduce using equation [6.1] that the forces acting on *bc* and *da* are equal in magnitude but in opposite directions along the x-axis and thus cancel to zero. The force exerted on *ab* is perpendicular to the x-axis, which is the direction of the current I within *ab*, and therefore lies in the (y, z) plane. The force is also perpendicular to the direction of $\tilde{B}$ and thus its direction makes an angle of θ with the y-axis and points below it. Using equation [6.1], we deduce that the amplitude of the force acting on the current-carrying conductor *ab* is $IB dx$. Using a similar argument, we deduce that the force acting on *cd* has the same amplitude but is directed in the opposite direction. The perpendicular separation between the lines of action of these two anti-parallel forces is $dy \cdot \sin(\theta)$, so that they together exert a couple $C(\theta)$ given by one force multiplied by

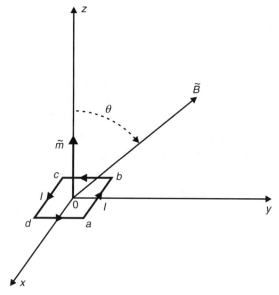

Fig. 6.3   A small square conducting loop *abcda* carries a current I and lies in the (x, y)-plane of three-dimensional Cartesian coordinates. The equivalent magnetic dipole moment of the loop $\tilde{m}$ points along the z-axis and the local magnetic field lies in the (y, z)-plane and makes an angle θ to the z-axis.

the perpendicular separation between the forces, as described in Section 4.2 for the electrical case. Thus

$$C(\theta) = IBdx[dy \cdot \sin(\theta)],$$

acting along the negative x-axis. Thus the couple $C(\theta)$ acting on a magnetic moment $m = Idxdy$ in a magnetic field $\tilde{B}$ when the angle between the moment and the field is θ, is given by

$$C(\theta) = (Idxdy)B \sin(\theta) = mB \sin(\theta). \qquad [6.6]$$

The couple acts along the direction perpendicular to both $\tilde{m}$ and $\tilde{B}$ and in the direction of advance of a right-hand thread rotated from the direction of $\tilde{m}$ to that of $\tilde{B}$. The couple acts such as to align the directions of $\tilde{m}$ and $\tilde{B}$.

Equation [6.6] has exactly the same form as equation [4.4] for the couple acting on an electric dipole in an electric field. The similarity of the equations describing an electric dipole in an electric field and a magnetic dipole in a magnetic field would lead us to expect that the energy $U_m(\theta)$ of a magnetic dipole $\tilde{m}$ pointing at an angle θ to the direction of a magnetic field $\tilde{B}$ will be given by

$$U_m(\theta) = -mB \cos(\theta), \qquad [6.7]$$

and this turn out to be true. However, we cannot deduce this result as we did in the electrical case without the artificial introduction of unit magnetic poles which we believe do not exist. An alternative method of deriving equation [6.7] is to calculate the work done by an external agency when the direction of a magnetic dipole is twisted from pointing along the direction of the magnetic field to pointing at an angle of θ to this field. To bring about such a rotation of the direction of the dipole the external agency must exert a couple equal in magnitude but opposite in direction to the couple $C(\theta)$ exerted by the magnetic field as in equation [6.6]. When the external couple $C(\theta)$ increases the angle between the directions of the magnetic dipole and the magnetic field from θ to $\theta + d\theta$, the small increment in the work done by the external agency dW is, by definition, given by $C(\theta)d\theta$ as described in Appendix 1. The total work $W(\theta)$ done when twisting the direction of the dipole from parallel to the magnetic field ($\theta = 0$) to making an angle θ with the field may be calculated by adding all the work increments dW by integration. The total work $W(\theta)$ must then be equal to the difference between $U_m(\theta)$ and $U_m(0)$:

$$U_m(\theta) - U_m(0) = \int_{\theta=0}^{\theta=\theta} dW = \int_0^\theta C(\theta)d\theta$$

$$= mB \int_0^\theta \sin(\theta)d\theta = -mB \cos(\theta) + mB.$$

This result is consistent with the conclusion that

$$U_m(\theta) = -mB\cos(\theta) \quad \text{and} \quad U_m(0) = -mB,$$

in agreement with equation [6.7]. As with the electric dipole, $U_m(0) = -mB$, $U_m(90°) = 0$ and $U_m(180°) = +mB$.

6.3 The magnetic field of a macroscopic current loop

Magnetic fields may be generated in the laboratory using macroscopic current loops consisting of a conductor carrying a current I in the form of a closed loop, as shown in Fig. 6.4. If we imagine any surface that is bounded by the conductor we could suppose that this surface is divided into a large number of very small areas or meshes by a number of extra conductors criss-crossing the surface and connected with the bounding conductor as shown in Fig. 6.4. Let us further imagine that a current I flows around the conductors that define each small meshes in the same direction as the current I that flows in the original macroscopic conductor. If we look at two neighbouring meshes we see that they share a common conductor and that the current circulating in the same direction around the two meshes must result in the same current flowing in opposite directions in the common conductor. Thus the currents circulating around all the

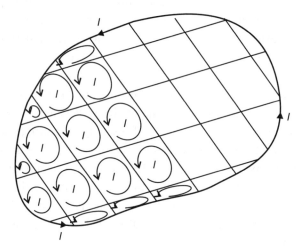

Fig. 6.4 A conducting loop of arbitrary shape carries a current I. On any surface bounded by the conducting loop, a mesh of conductors can be imagined such that every small mesh carries a current I circulating in the same direction as the current in the original conducting loop. Because of the cancellation between the currents carried by the conductor that is shared in common between two adjacent meshes, the current in all the added imagined meshes is zero. Thus the magnetic field created by the original current-carrying loop is exactly equivalent to that which would be created by summing the effects of all the imaginary current-carrying meshes.

imaginary conductors that we added to the original current loop cancel to zero and the only non-zero current remaining is that flowing in the original peripheral conductor. This demonstrates that the sum of all the currents I flowing around each of the small meshes is equivalent to the single original current I flowing in the original macroscopic circuit. We may then calculate the field generated by the macroscopic circuit by summing the magnetic fields generated by the magnetic moments created by all the small meshes around which circulates the current I, using equations [6.3] and [6.4].

6.4   The magnetostatic potential of a magnetic dipole

If we choose, we may derive a magnetostatic potential which is represented by the symbol ϕ and which takes the place of the electrostatic potential or voltage V. The relationship used to relate the magnetic field to the magnetostatic potential, and which defines the magnetic potential, is slightly different from that in the electrical case shown in equation [1.10]:

$$B_x = -\mu_0 \frac{d\phi}{dx}, \quad B_y = -\mu_0 \frac{d\phi}{dy}, \quad B_z = -\mu_0 \frac{d\phi}{dz}.$$

In general, the magnetic field component B_R in a direction $\tilde{R}$ is given by

$$B_R = -\mu_0 \frac{d\phi}{dR} \quad \text{or} \quad d\phi = -\frac{1}{\mu_0} B_R dR, \qquad [6.8]$$

where $d\phi$ is the small change in the magnetostatic potential ϕ when we move a small distance dR in the direction of $\tilde{R}$. Note that the magnetic field components are defined as μ_0 times the negative gradient of the potential rather than simply as the negative gradient of the potential as in the electrical case. Using equations [6.8], we may deduce the total change in the magnetostatic potential ϕ when we move between two points with coordinates represented by a and b along a specified path by summing the incremental changes $d\phi$ for each step dR, using a line integral as follows:

$$\phi(b) - \phi(a) = \int_a^b d\phi = -\frac{1}{\mu_0} \int_a^b B_R dR. \qquad [6.9]$$

The same procedure was used in the electrostatic case to deduce equation [1.11]. In the electrostatic case it can be proved that the change in the electrostatic potential or voltage between two given points is the same no matter what path is chosen between them. As we shall see below in Section 6.5, the same is true for the magnetostatic potential, with the additional proviso that the path taken does not thread any of the current loops that create the magnetic field.

For the two-dimensional polar coordinates (see Appendix 1) used in Fig. 6.2b we can deduce the relation between the components of the magnetic field B_R and B_θ and the magnetostatic potential $\phi(R, \theta)$. When R changes incrementally to $R + dR$ with θ held constant, the field point $P(R, \theta)$ moves a distance dR so that the change in ϕ becomes $d\phi = -(1/\mu_0)B_R dR$ so that, as in equation [6.8], we have

$$B_R(R, \theta) = -\mu_0 \left(\frac{d\phi}{dR}\right). \qquad [6.10]$$

Similarly, when θ changes incrementally to $\theta + d\theta$ with R held constant, the field point $P(R, \theta)$ moves a distance $R \cdot d\theta$ along the circumference of a circle of radius R. Using equation [6.8], we can write the resultant change in ϕ as $d\phi = -(1/\mu_0)B_\theta(Rd\theta)$, so that we have

$$B_\theta(R, \theta) = -\mu_0 \frac{1}{R} \left(\frac{d\phi}{d\theta}\right). \qquad [6.11]$$

The expression for the magnetostatic potential ϕ_D of a magnetic dipole $\tilde{m}$ may be deduced from a comparison of equations [6.10] and [6.11] with the experimentally determined magnetic field components of a magnetic dipole given in equation [6.4]. It is given by

$$\phi_D = \frac{m \cos(\theta)}{4\pi R^2}. \qquad [6.12]$$

As we would expect, this has the same R- and θ-dependence as the electric potential of an electric dipole as shown in equation [4.1] but, unlike the electric potential, which contains ε_0, it does not contain the magnetic constant μ_0. Instead this magnetic constant occurs in equations such as [6.8] and [6.9] which connect the magnetic potential to the magnetic field.

6.5. The magnetostatic potential near a current loop and Ampère's law

We will now calculate the magnetostatic potential in the region close to a macroscopic current loop. First we divide the area of the loop into small meshes as in Fig. 6.4. Each mesh has an area dA and a current of I circulating around it so that it has a magnetic moment of $m = dAI$ perpendicular to its plane. Applying equation [6.12], we deduce that the magnetostatic potential of this single mesh ϕ_M at a field point $P(R, \theta)$ is given by

$$\phi_M = \frac{dAI \cos(\theta)}{4\pi R^2}. \qquad [6.13]$$

where R and θ are the polar coordinates of the field point $P(R, \theta)$ relative to the position and direction of the dipole moment of the mesh as shown in Fig. 6.2b. The potential for the whole loop ϕ_L is then calculated by summing over all the small meshes. Figure 6.5a depicts a section through a plane macroscopic current loop seen from the side, which shows the magnetic field generated by the macroscopic current loop as closed loops threading the circuit as we would expect. At a field point remote from the loop, like that marked as a in Fig. 6.5a, the magnetostatic potential is determined by the summation of the potentials of the individual meshes each represented by an expression like that in equation [6.13]. Let the total summed potential at a generated by the macroscopic current loop be written as $\phi_L(a)$. If we move in any closed path like that sketched in the diagram which starts and ends at the point a, we expect that we will obtain the same value of $\phi_L(a)$, the potential at a, at the start and end of tracing out the

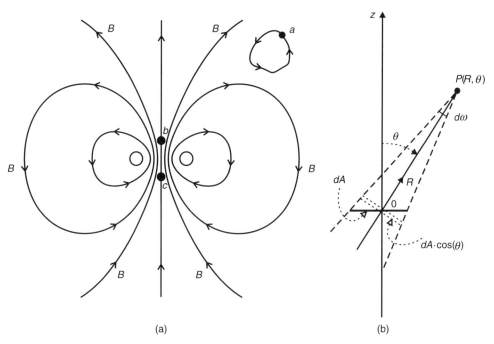

(a)   (b)

Fig. 6.5   (a) A section through a plane macroscopic current-carrying loop of conductor, indicated by two open circles, with the magnetic field $\tilde{B}$ created by the loop in the plane of the section. An arbitrary circuit remote from the current carrier is shown at a. Two points b and c in this plane are located just above and just below the plane of the current-carrying loop. (b) A small plane area dA of the plane macroscopic current loop shown perpendicular to the z-axis. The field point $P(R, \theta)$ is a distance R away from the centre of the area dA, such that $\tilde{R}$ makes an angle θ to the z-axis. The elemental solid angle subtended by dA at $P(R, \theta)$ is shown as $d\omega$. The solid angle $d\omega$ is equal to the projection $dA \cos(\theta)$ of the area dA perpendicular to the direction of $\tilde{R}$ divided by R^2 so that $d\omega = dA \cos(\theta)/R^2$.

closed path. Using equation [6.9], we may deduce that the line integral of $\tilde{B}$ around this closed path is zero for all such paths:

$$\oint B_R dR = \int_a^a B_R dR = -\mu_0[\phi_L(a) - \phi_L(a)] = 0. \qquad [6.14]$$

This is the equivalent to equation [1.12] for the line integral of the electric field. The explanation for this relation is that in a circuit, such as that including the point a in the figure, at some parts of the path the component B_R of the field $\tilde{B}$ in the direction of the path is positive (potential falling), and at others it is negative (potential rising), so that in a complete circuit the potential change and hence the line integral of the field around that path is zero.

However, we can choose a path which starts from the point marked b in the figure, just above the loop, and ends at the point marked c, just below the loop, such that the component of the magnetic field is always positive along the path. In this case we expect the potential to fall along the whole path. Because we can make the points b and c as close together as we like, so that the line integral of the field $\tilde{B}$ between c and b becomes negligible, it must be true that the line integral of $\tilde{B}$ evaluated around a closed loop starting and ending at the point marked b or c is negative and non-zero. What distinguishes these paths from those used to deduce equation [6.14] is that they thread through the magnetic field-generating current circuit. We will first calculate the drop in the potential along such a path and derive a general relationship that replaces equation [6.14] and which holds for any path. In Section 6.6 we will show why this magnetic case differs from the electrostatic case in which the line integral of $\tilde{E}$ around any closed loop is always zero.

If we rearrange the terms in equation [6.13], we can write the magnetostatic potential of a single mesh at the field point $P(R, \theta)$ as sketched in Fig. 6.5b as

$$\phi_M = \left[\frac{dA \cos(\theta)}{R^2}\right]\left(\frac{I}{4\pi}\right). \qquad [6.15]$$

As described below, the square bracket in this expression may be identified as the solid angle subtended at a general field point $P(R, \theta)$ by the area dA of the elementary mesh.

The familiar **plane angle** is the angle subtended at a point by a line. The size of the angle that is subtended at the centre of a plane circle of radius R, by a length L of the circumference of the circle, is defined in radians to be L/R. Because the complete circumference has a length $2\pi R$, the complete plane angle subtended at the centre is equal to 2π radians. In contrast, a **solid angle** is a conical region of space subtended at a point by an area. The magnitude of a solid angle that is subtended at the centre of a sphere of radius R by an area A of the surface of that sphere is defined to be A/R^2. The units for a solid angle are steradians, and because the surface area of a sphere is given by $4\pi R^2$, the complete solid angle is 4π.

Returning to the magnetostatic potential of the small mesh in equation [6.15], if we construct a sphere centred on the field point $P(R, \theta)$ as in Fig. 6.5b with a radius R, then $dA \cdot \cos(\theta)$ is the area of the mesh dA projected on to the surface of the sphere and the first square bracket in equation [6.15] is the solid angle subtended at the field point P by the mesh area dA. We need not worry about the difference in area of the spherical surface which is curved and the projected area which is flat, if dA is infinitesimal in area. If the elemental solid angle is denoted by $d\omega$, the magnetostatic potential generated by the current-carrying mesh of area dA at the field point P in Fig. 6.5b may be written as

$$\phi_M(R, \theta) = \left(\frac{d\omega}{4\pi}\right) I.$$ [6.16]

At the point marked b in Fig. 6.5a, just above the loop, the magnetostatic potential of the whole loop ϕ_L may be found by summing the potentials like that in equation [6.16] due to all the small meshes. This requires the sum of all the elementary solid angles $d\omega$ subtended at the field point b by all the meshes. As the point b approaches the plane of the loop, the total solid angle subtended at this point by the plane of the loop clearly approaches 2π, half the total possible solid angle. In a similar manner, the total solid angle subtended by the macroscopic current loop at the corresponding point marked c in Fig. 6.5a, just below the loop, may be deduced to be -2π. The negative sign of this solid angle has its origin in the fact that calculating the solid angle subtended at the point marked b of a given mesh involves $\cos(\theta)$ whereas the calculation of the solid angle subtended at the point marked c by the same mesh involves $\cos(-\theta) = -\cos(\theta)$. Thus we have deduced that the change in the magnetostatic potential of the whole current loop on moving from the point marked b just above the plane of the current loop to the point marked c just below the loop is given by the expression

$$\phi_L(c) - \phi_L(b) = -\frac{I}{4\pi}(2\pi + 2\pi) = -I.$$ [6.17]

Because we can make the points b and c as close together as we like so that the line integral of $\tilde{B}$ between them is negligible, we may finally deduce, using equation [6.9], that for any paths that thread the current circuit

$$\oint B_R dR = -\mu_0 \oint d\phi_L = +\mu_0 I.$$ [6.18]

This law, called **Ampère's law**, is the single most important equation in magneto-statics and, together with Gauss's law (equation [6.5]), it is capable of predicting all of magnetostatics. In words, it states that **the line integral of the magnetic field $\tilde{B}$ around any closed path is equal to μ_0 multiplied by the total current that threads that path**. Note that saying that the integration path is threaded by the current is equivalent to saying that the current circuit is threaded by the chosen integration path. There is a sign involved in equation [6.18]. We derived it for a path that threads the current circuit in the direction of the magnetic moment of the circuit, or in the direction of advance of a right-handed thread rotated in the same sense as the current circulates around the circuit. The line integral of $\tilde{B}$ around a closed path that threaded the current circuit in the opposite direction would be equal to $-\mu_0 I$. Equation [6.14] may now be seen as a particular case of this law when the path chosen does not thread the current circuit and thus no current I threads the chosen path.

Ampère's law depends upon the current that threads the integration path. If there were several different current-carrying circuits threading the integration path, then the proof above could be applied to each current circuit in turn and the line integral of the magnetic field would be related to the total current threading the path. Taking this process to the limit, we may state Ampère's law when the current is distributed in space. The total current threading the integration path is then related to the total flux of current density $\tilde{J}$ (current per unit area) that cuts any surface that is bounded by the integration path by the equation

$$\oint B_R dR = \mu_0 \iint_S J_N dS. \tag{6.19}$$

The line integral on the left-hand side of equation [6.19] is taken over a definite chosen path. Let us now imagine a particular surface that is bounded by the chosen integration path. The surface integral on the right-hand side of equation [6.19] calculates the total flux of current passing through that surface. Within the surface integral, J_N is the current per unit area flowing normally to the an infinitesimal area dS of the chosen surface so that the current flux through dS is $J_N dS$. The surface integral performs the summation of $J_N dS$ over areas dS such that these elemental areas cover the chosen surface exactly once.

6.6 The differences between the electrostatic and magnetostatic potentials

Ampère's law illustrates one major difference between electrostatics and magneto-statics. Because the line integral of $\tilde{B}$ around a closed path that threads a current circuit is non-zero, it follows from equation [6.9] that the magnetostatic potential is no longer single-valued at a given point. For this reason the usefulness of the magnetostatic potential is limited to cases in which we are dealing with magnetic fields at positions that are remote from the current loops that create the fields. To illustrate the difference with the electrostatic case where the electrostatic potential or voltage is single-valued, we will construct an electrostatic dipole layer that resembles the magnetic dipole layer that is formed by a current loop.

In Fig. 6.6a we show a flat, parallel-sided, sheet of thickness t which bears a positive charge on its upper surface and an equal negative charge on its lower surface. If the surface charge per unit area on the upper and lower surfaces is given by $+\sigma_S$ and $-\sigma_S$ coulombs per square metre then the sheet has an electric dipole moment per unit area of $P_A = \sigma_S t$ coulombs per metre. The electric field external to the sheet is seen to originate at the positive charge on the top surface of the sheet and end on the negative charge on the lower surface of the sheet. If we choose a small area dS of the sheet, this has an electric dipole moment of $p = P_A dS$ and the voltage $V(R, \theta)$ due to this dipole moment at a field position $P(R, \theta)$ with polar coordinates of R and θ relative

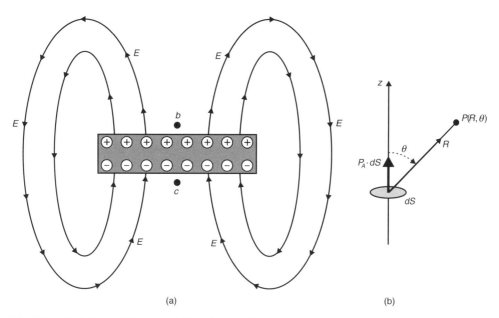

(a)   (b)

Fig. 6.6   (a) A plane slab of material bearing a positive surface charge density on its upper face and an equal negative surface charge density on its lower face. The resulting electric field in the plane of the diagram is indicated by $\tilde{E}$. A position just above the plane of the slab is indicated as b and a position just below the slab is indicated by c. (b) A small area of the charged slab dS is shown perpendicular to the z-axis. The field point $P(R, \theta)$ is shown a distance R from the centre of dS such that the direction of $\tilde{R}$ makes an angle θ with the z-axis. If the dipole moment per unit area of the slab is given by P_A then the electric dipole moment of the small area dS of the slab is $P_A dS$.

to this moment, as shown in Fig. 6.6b, is given by equation [4.3] with $P_A dS$ substituted for p:

$$V(R, \theta) = \frac{P_A dS \cos(\theta)}{4\pi\varepsilon_0 R^2}$$

$$= \left(\frac{dS \cos(\theta)}{R^2}\right)\left(\frac{1}{4\pi}\right)\left(\frac{P_A}{\varepsilon_0}\right) = \left(\frac{d\omega}{4\pi}\right)\left(\frac{P_A}{\varepsilon_0}\right), \qquad [6.20]$$

where $d\omega$ is the solid angle subtended at the field point by the small area dS. This has exactly the same form as equation [6.16] for the magnetic loop with (P_A/ε_0) substituted for I. Following exactly the same argument as used in the magnetic case considered above, we can deduce that the drop in voltage along an external path that starts at a point marked b, just above the dipole sheet as shown in Fig. 6.5a, and ends at the point marked c, just below the sheet, is given by

$$V(c) - V(b) = -\frac{P_A}{\varepsilon_0}. \qquad [6.21]$$

The major difference between the electrical and the magnetic case is seen when we complete the closed path in the electrical case and calculate the change in V on moving from c to b across the sheet. Within the sheet there is a strong electric field which is directed downward in Fig. 6.6a from the positively charged upper surface towards the negatively charged lower surface of the sheet. Thus on passing through the sheet from c to b the voltage rises. Let us construct a short circular cylinder with its axis perpendicular to the sheet which encloses some of the positive charge on the upper surface of the sheet. If we apply Gauss's law to this cylinder we deduce that an electric field E_S **leaves** each flat end of the cylinder, where

$$2E_S = \frac{\sigma_S}{\varepsilon_0} \quad \text{or} \quad E_S = \frac{\sigma_S}{2\varepsilon_0}. \qquad [6.22]$$

We have used the symmetry of a plane charged layer to deduce that the electric field at each end of the Gauss law cylinder has the same amplitude. By constructing a similar Gauss law cylinder enclosing some of the negative charge of density $-\sigma_S$ on the lower face of the dipole layer, we would deduce that an electric field of the same amplitude E_S **enters** each flat end of that cylinder. Thus the total downward-pointing electric field within the dipole layer depicted in Fig. 6.6a is given by

$$E_L = 2E_S = \frac{\sigma_S}{\varepsilon_0}. \qquad [6.23]$$

Because the electric field is constant within the dipole layer, the change in the voltage $V(b) - V(c)$ on moving upward through the layer is given by

$$V(b) - V(c) = +E_L t = +\frac{\sigma_S}{\varepsilon_0} t = \frac{+P_A}{\varepsilon_0}.$$ [6.24]

Thus the fall in voltage between the positions b and c along a path external to the sheet is exactly cancelled by the rise in voltage between the positions c and b passing through the sheet so that the voltage change round the complete circuit is zero. Thus the line integral of $\tilde{E}$ around any closed path is zero whether the path penetrates a dipolar layer or not. This is enough to ensure that the electrostatic potential or voltage is single-valued, as discussed in Section 1.6.

If we return to the comparison between the electrostatic and magnetostatic potentials, we see that the changes in the two potentials are similar external to the dipole layer or current loop, but in the magnetic case there is no oppositely directed field between c and b to return the potential to its initial value on completing the closed path bcb. To sum up the difference between the electrical and the magnetic case, we repeat the equation used for electrostatics as equation [6.25] and that for magnetostatics as equation [6.26]:

$$\oint E_R dR = 0,$$ [6.25]

$$\oint B_R dR = \mu_0 \iint_S J_N dS.$$ [6.26]

Topics covered in Chapter 6

1.   Just as electrical charges create electric fields, current-carrying conductors and magnetic dipoles or equivalently current loops create magnetic fields. There is no magnetic charge that could take the place of electrical charge.

2.   The magnetic field around a magnetic dipole forms closed loops and has the same shape as that of the electric field around an electric dipole. Both types of field fall off with the inverse cube of the distance from the dipole. However, the dependence of the amplitude of the field on the amplitude of the dipole moment is different in electricity and in magnetism.

3.   A magnetic dipole is formed when current flows around a small plane loop. The magnetic dipole moment associated with the small loop is the product of its area and the amplitude of the current which flows around it. It is directed perpendicularly to the plane of the small loop and points in the direction of advance of a right-hand thread rotated in the direction of the circulating current flow.

4.   The magnetic field created by a macroscopic current loop carrying a current I constructed in the laboratory may be calculated by dividing the macroscopic loop into a very large number of very small current loops or meshes, each carrying a circulating current I and summing the contributions of all the small loops. The currents flowing in the small loops can be made to cancel to zero everywhere except around the original macroscopic loop so that the current distributions in the two cases are physically equivalent.

5.   A magnetic dipole experiences no translational force in a uniform magnetic field. It does, however, experience a couple which tends to align the magnetic dipole moment with the direction of the local magnetic field.

6.   Because the magnetic field of a magnetic dipole forms closed loops, Gauss's law in magnetism states that the total outward flux of the magnetic field through a closed surface of any shape is always zero.

7.   Ampère's law states that the line integral of the magnetic field around any closed loop is equal to the total current that threads through that loop multiplied by a numerical constant.

8.   A magnetostatic potential may be defined in magnetism in a manner similar to the electrostatic potential or voltage in electricity. However, the magnetostatic potential is of much less importance because at a single location, near current loops, the magnetic potential may take many values. Far from the current loops that create the magnetic field, the magnetic potential is a single-valued function of the position like the voltage created by static electric charges.

Important equations of Chapter 6

1.   The force on a short length dL of conductor carrying a current $\tilde{I}$ in a magnetic field $\tilde{B}$ has magnitude given by

$$dF = IBdL \sin(\theta) = |(\tilde{I} \times \tilde{B})|dL.$$

The force acts at right angles to the directions of both $\tilde{I}$ and $\tilde{B}$ and points in the direction of advance of a right-hand thread rotated from the direction of $\tilde{I}$ to that of $\tilde{B}$. The angle θ is that between the directions of $\tilde{I}$ and $\tilde{B}$.

2.   The magnitude of the magnetic dipole moment $\tilde{m}$ due to a small loop of conductor carrying a current I which encloses an area dA is given by

$$m = IdA.$$

The moment points normal to the loop and in the direction of advance of a right-hand thread rotated in the direction of the circulating current.

3.   The magnetic field components $B_R(R, \theta)$ and $B_\theta(R, \theta)$ generated by a magnetic moment $\tilde{m}$ which point along the directions of increase of R and θ at the field point $P(R, \theta)$, as in Fig. 6.2b, are given by

$$B_R(R, \theta) = \frac{2\mu_0 m \cos(\theta)}{4\pi R^3} \quad \text{and} \quad B_\theta(R, \theta) = \frac{\mu_0 m \sin(\theta)}{4\pi R^3}.$$

4.   A magnetic moment $\tilde{m}$ which is oriented at an angle θ to the direction of a magnetic field $\tilde{B}$ has an energy $U(\theta)$ and is acted on by a couple $C(\theta)$ such that

$$U(\theta) = -mB \cos(\theta) \quad \text{and} \quad C(\theta) = +mB \sin(\theta).$$

5.   Gauss's law states that the total flux of $\tilde{B}$ through any closed surface is zero so that

$$\iint_S B_N dS = 0.$$

6.   Ampère's law states that the line integral of $\tilde{B}$ around any closed path is equal to the total flux of current that penetrates any surface bounded by the path:

$$\oint B_R dR = \mu_0 \iint_S J_N dS.$$

J_N is the current per unit area flowing normally through an element of surface area dS of any surface bounded by the path over which the line integral is calculated.

Problems

6.1   The difference in voltage between two terminals located close together, feeds a constant current I through a length of flexible wire which lies on a smooth plane, parallel to the (x, y) plane. When a large magnetic field is applied along the direction of the z-axis the wire forms a circle. Explain this phenomenon.

6.2   A magnetic dipole $\tilde{m}$ is constrained to point along the x-axis. A magnetic field is applied with a component along the x-direction B_x which has an amplitude which decreases as x increases. Prove that the field exerts a translational force on the dipole. In what direction does the force act?

6.3   Two small plane circular loops of wire, centred on the z-axis, are located in planes parallel to the (x, y) plane. Assume that the separation of the coils along the z-axis is much greater than the radius of either coil. If a current flows around each coil in the same sense, show that the coils attract each other with a force directed along the z-axis (a) by considering the magnetic dipole moment of one coil in the field due to the other and (b) by considering directly the effect of the magnetic field of one coil acting upon the current flowing in the second coil as predicted by equation [6.1].

6.4   A magnetic dipole with a magnetic moment m is located at the origin of coordinates 0 and is aligned along the z-axis. Calculate the z-component of the magnetic field of the dipole at the field point $P(R, \theta)$ where the polar coordinates R and θ relative to the origin are defined in Fig. 6.2b. Show that at all points $P(R, \theta)$ with an angle θ_1 such that $\cos^2(\theta_1) = 1/3$, the z-component is zero. Hence show that the magnetic interaction energy of two dipoles pointing in the same direction is negative only if the second dipole is located within a circular cone of semi-angle θ_1 with its apex at the first dipole and with the cone axis parallel to the dipole moments.

6.5   A magnetic dipole with magnetic moment m_1 is situated at the origin of the polar coordinates shown in Fig. 6.2b and points along the z-axis. A second dipole of strength m_2 is centred at the point $P(R, \theta_1)$ and its moment makes an angle θ_2 with the positive direction of the radius vector R from the origin. If the two dipoles lie in the same plane, use equations [6.4] and [6.7] to obtain an expression for the magnetic energy of interaction of the two dipoles. For a fixed value of R, find the values of θ_1 and θ_2 that (a) maximize and (b) minimize this interaction energy. If $\theta_1 = \pi/2$ and R is fixed, what values of θ_2 maximize and minimize the interaction energy?

6.6   A small plane circular coil of radius a carries a current I. Find the direction and estimate the strength of the magnetic field at a position in the plane of the coil and a distance R away from the centre of the coil, where $R \gg a$. Justify the assumptions that you made.

7. The generation of magnetic fields

The magnetic fields that are created by the currents carried by the straight wire, the long solenoid and the circular plane coil are deduced using only Gauss's law and Ampère's law. The Biot–Savart law, which can be used to calculate the magnetic fields generated by currents flowing in a conductor of any geometry, is introduced. The regions where the magnetic field varies by less than 1% and by less than 5% are calculated for the most practical configurations of coils used to generate magnetic fields in the laboratory.

7.1 The magnetic field around a straight current-carrying wire

In the previous chapter we discussed the form of magnetic fields that are generated by current loops. In this chapter we will consider currents that have different geometries and calculate the magnetic fields they create. When the geometry of the current has high symmetry we can often calculate the magnetic field using only Ampère's and Gauss's laws. The simplest example is a long straight wire carrying a current of I pointing along the z-axis, as shown in Fig. 7.1a. We will calculate the magnetic field at a point P a distance R from the wire. With no loss of generality we can chose the point P to be on the x-axis, as sketched in Fig. 7.1a. Consistent with the cylindrical symmetry of this problem, the magnetic field at P could have components B_z in the z-direction, B_R pointing radially away from the wire in the x-direction or B_T in a tangential direction circling around the wire, in the y-direction at P. The symmetry of this problem also demands that while these field components may depend upon the distance from the wire R they may not depend on the distance z along the wire for a long wire. We will show that B_z and B_R are zero and we will derive an expression for B_T.

In Fig. 7.1b we show a rectangular path $abcda$ located in the (x, z) plane with sides ab and cd parallel to the wire and sides bc and ad in the x-direction. From Ampère's law the line integral of $\tilde{B}$ around this path is zero as no current threads through the integration path. The contribution to the integral from the paths bc and da must cancel to zero as they are traversed in opposite directions. The magnetic field caused by the wire in the vicinity of the path cd may be made as small as we wish by extending the lengths of bc and da so that the line integral of $\tilde{B}$ around the complete path $abcda$

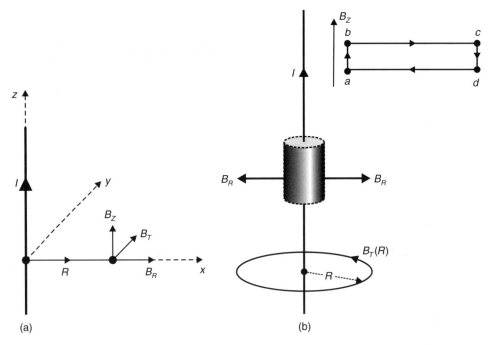

Fig. 7.1   (a) A straight wire carries a current I which is directed along the z-axis. At a point in the (x, z)-plane a distance R from the wire the three possible directions for the magnetic field generated by the wire are B_R, B_T and B_z which are directed along the x-, y- and z-axes, respectively. (b) The straight wire again conducts a current I in the z-direction. A rectangular circuit $abcda$ in the plane of the diagram and a circular circuit of radius R in a plane normal to the wire are shown. Also shown is a cylindrical Gauss law circuit which is co-axial with the wire.

reduces simply to $B_z ab$. Thus we have shown that $B_z ab$ is zero while ab is finite, which ensures that B_z is zero. Because we can place the circuit where we wish, $B_z = 0$ at all radial distances R from the wire. If we construct a Gauss law cylindrical surface about the wire, as shown in Fig. 7.1b, we can immediately say that $B_R = 0$ as the total flux of $\tilde{B}$ out of the cylinder is zero by Gauss's law and we have already shown that B_z is zero. The component B_T is tangential to the surface of the Gauss law cylinder and does not contribute to the flux of magnetic field out of the cylinder.

Figure 7.1b shows the circular path along which B_T will point as we move around the wire at a distance R from it. By the cylindrical symmetry of the problem, the value of $B_T(R)$ is constant around this path and points along it, so that the line integral of $\tilde{B}$ around this path is equal to $2\pi R B_T$. The circular path is threaded by a current I, and by Ampère's law we may deduce that

$$2\pi R B_T = \mu_0 I \quad \text{or} \quad B_T(R) = \frac{\mu_0 I}{2\pi R}. \quad\quad [7.1]$$

We have thus shown that the direction of the magnetic field created by a straight conductor carrying a current I forms circles about the axis of the wire, such that the circles lie in a plane perpendicular to the wire. The amplitude of this field is proportional to the current I and inversely proportional to the radius R of the circle. The field circulates in a direction that will cause a right-handed thread to advance in the direction of the current in the central wire.

7.2   The long solenoid

Another important geometry is that of a long coil wound on a cylindrical former of radius a using wire carrying a current I such that there are N turns of wire per unit length of the coil. Such a coil is called a **solenoid**, and a longitudinal section through such a coil is shown in Fig. 7.2. This configuration has the same cylindrical symmetry as the straight wire so that the magnetic field at the field point P a distance R from the axis of the coil could have components B_z, B_T and B_R. Again by symmetry for a sufficiently long coil, the values of these field components, far from the ends of the coil, may not depend on the value of z, the coordinate along the cylinder axis as shown in the figure.

That B_R is zero both inside and outside the solenoid may be seen by applying Gauss's law to closed cylindrical surfaces with the same axes as the coil and having radii less than and greater than a. The flux due to B_z **in** through one flat end of such a cylindrical Gauss law surface must be matched by the same flux of B_z **out** of the other flat end as B_z may not vary with z. The fact that the total flux of B_R out of the curved surfaces of these Gauss law cylinders is zero requires that B_R is zero inside and outside the solenoid.

If we construct a circular path centred on the axis of the coil and perpendicular to it and which has a radius R_1 within the coil ($R_1 < a$), we may deduce that the line integral of $B_T(R_1)$ around this path is $2\pi R_1 B_T(R_1)$. Because no current threads this path, we may deduce using Ampère's law that everywhere inside the coil $B_T(R_1) = 0$. If we construct a similar circular path with radius R_2 outside the coil ($R_2 > a$), we again deduce that the line integral of $B_T(R_2)$ around this path is $2\pi R_2 B_T(R_2)$; but now, because of the spiral nature of the coil, there is a total current of I threading this plane circuit so that we obtain

$$B_T(R) = \frac{\mu_0 I}{2\pi R} \quad \text{for} \quad R > a. \tag{7.2}$$

Finally, we construct a rectangular path like that shown as $abcda$ in Fig. 7.2 such that the sides ab and cd have a length L. Because the contributions to the line integral of $\tilde{B}$ along bc and da are zero as $B_R = 0$ and that along cd may be made arbitrarily small by

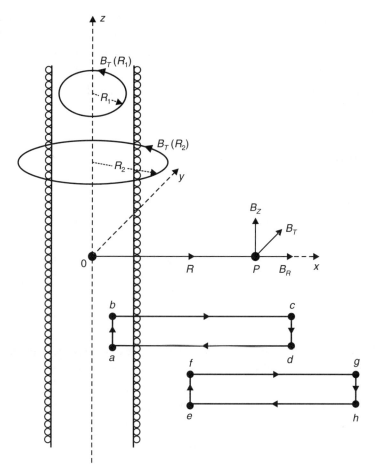

Fig. 7.2   A section through a long cylindrical solenoid with its axis directed along the z-axis which carries a circulating current I. At a point P in the (x, z)-plane, a distance R from the axis of the solenoid, are shown the three possible components of the magnetic field generated by the current in the solenoid, B_R, B_T and B_z, which are directed along the x-, y- and z-axes. Two rectangular circuits in the plane of the diagram, *abcda* and *efghe*, are shown together with two circles of radii R_1 and R_2 in planes normal to the axis of the solenoid.

lengthening *bc* and *da*, the only contribution to the line integral is $B_z L$ which comes from the part of the path *ab*. The total current threading the path *abcda* is NIL, so that we obtain from Ampère's law the relation

$$B_z(R) = \mu_0 NI \quad \text{for} \quad R < a. \tag{7.3}$$

From the fact that we would obtain this result for any position of the path *ab* inside the solenoid, we may deduce that the field B_z is constant within the solenoid and does not, for example, depend upon R. If we apply the same argument to the circuit shown as *efghe* in Fig. 7.2, which is threaded by no current, we may deduce that

$$B_z(R) = 0 \quad \text{for} \quad R > a. \tag{7.4}$$

Using these arguments, we have deduced that the field inside a long solenoid B_z has a constant amplitude of $\mu_0 NI$, where N is the number of turns per unit length of the solenoid and I is the current carried by the wire, and it is directed along the axis of the solenoid. Outside the solenoid the only field B_T has a tangential direction circling around the axis of the coil and has the same amplitude as would be generated by a straight wire carrying a current I along the axis of the coil.

The solenoid is described as long and the results obtained above hold in regions remote from the ends of the solenoid. We would expect that the field at the ends of a finite-length solenoid will drop, and this is the case. Also, because the solenoid may be considered as equivalent to a long chain of magnetic dipoles created by the successive current loops, we would expect the field to splay out at the ends and no longer to be axial. One simple deduction is that at the end of a long solenoid and **on its axis**, the field will be axial and of amplitude $\frac{1}{2}\mu_0 NI$. The axial direction of this field is a result of the axial symmetry of the solenoid. The amplitude may be deduced from the fact that two such solenoids placed end to end, having the same radius and number of turns per unit length, wound in the same sense and carrying the same current, must be equivalent to a single long solenoid of twice the length. Thus each of the solenoids must produce a field on the axis and at its end equal to one-half the field found within a long solenoid far from its ends.

7.3   The Biot–Savart law

We have deduced the magnetic field around a straight wire and the long solenoid using only the Gauss and Ampère laws partly to illustrate the application of these laws. For current configurations with lower symmetry this can become a tedious procedure and it is possible to deduce from the two fundamental laws a general relation between an arbitrary configuration of current and the field it will produce, called the **Biot–Savart law**. In Fig. 7.3a we show a short section of a conductor of length dL carrying a current I in the z-direction. At a field point $P(R, \theta)$ defined by the polar coordinates R and θ relative to the small length of conductor, the contribution $dB(R, \theta)$ to the magnetic field from this small current-carrying section is given by the Biot–Savart law as

$$dB(R, \theta) = \frac{\mu_0 I dL \sin(\theta)}{4\pi R^2}. \tag{7.5}$$

The angle θ is that between the directions of the current element dL and the radius vector R from the current element to the field point P. The field component $dB(R, \theta)$ is

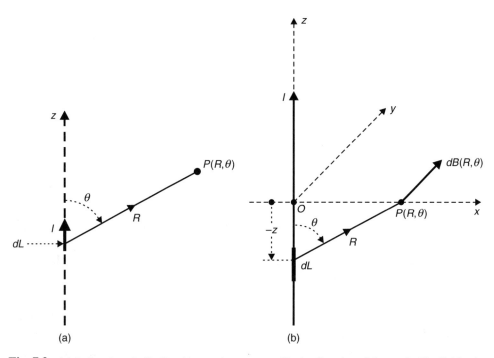

Fig. 7.3   (a) A short length dL of a wire carrying a current I in the direction of the z-axis. The field point $P(R, \theta)$ is a distance R from dL such that the angle between the directions of $\tilde{R}$ and the z-axis is θ. (b) A wire carrying a current I is directed along the z-axis of a Cartesian set of axes. A short length dL of the wire is located a distance z below the origin and on the z-axis. A field point $P(R, \theta)$ is located on the x-axis a distance R from dL and such that the angle between the directions of $\tilde{R}$ and the z-axis is θ. The Biot–Savart law predicts that the magnetic field at $P(R, \theta)$ caused by the current in the short length of conductor dL is $dB(R, \theta)$ and is directed along the y-axis.

perpendicular to both the direction of $\tilde{I}$ and of the vector $\tilde{R}$, and it points in the direction of advance of a right-handed thread rotated from the direction of I to the direction of R. With no lack of generality we may choose our axes such that the point P lies in the (x, z) plane as in Fig. 7.3b, and in this case the component of the magnetic field $dB(R, \theta)$ points along the y-axis. To obtain the value of the total magnetic field at the field point P the magnetic field contributions of all the elements of length dL along the current-carrying conductor must be summed. This can be achieved by integrating the expression in equation [7.5] along the whole length of the current-carrying conductor as we will illustrate below for a straight conductor. Using the compact notation of the vector product introduced for equation [6.2] (see Appendix 1) the Biot–Savart law my be written compactly as

$$d\tilde{B}(R, \theta) = \frac{\mu_0 dL}{4\pi R^3} [\tilde{I} \times \tilde{R}].$$
[7.6]

where the square bracket represents the vector product of the two vectors $\tilde{I}$ and $\tilde{R}$ which has an amplitude $IR \sin(\theta)$.

To illustrate the use of equation [7.5], we will calculate again the magnetic field due to a straight conductor directed along the z-axis and carrying a current I as shown in Fig. 7.3b. A typical small section of the conductor dL, or equivalently dz in these coordinates, is on the z-axis a distance z away from the origin of the coordinates at O. The field point P is taken as on the x-axis a distance $OP = R_0$ from the wire. The direction of the field increment $dB(R, \theta)$ at P is in the direction of the y-axis for all locations of the section dL of the current-carrying straight conductor, in agreement with our conclusions in Section 7.1. To find the total field we must integrate the expression on the right of equation [7.5] over all values of z from $-\infty$ to $+\infty$. Noting that $\sin(\theta) = R_0/R$ and that $R^2 = R_0^2 + z^2$ the expression for the total tangential magnetic field $B_T(R_0)$ at P becomes

$$B_T(R_0) = \int_{z=-\infty}^{z=+\infty} dB(R, \theta) = \int_{z=-\infty}^{z=+\infty} \left(\frac{\mu_0 I}{4\pi}\right)\left(\frac{\sin(\theta)}{R^2}\right) dz,$$

$$= \frac{\mu_0 I}{4\pi} \int_{-\infty}^{+\infty} \frac{R_0 dz}{(R_0^2 + z^2)^{3/2}} = \frac{\mu_0 I}{2\pi R_0} \qquad [7.7]$$

in agreement with equation [7.1].

7.4   The single coil and the Helmholtz pair of coils

Other than the straight wire and the long solenoid, several other configurations have proved useful for the generation of magnetic fields in the laboratory. One is the plane circular coil of radius R consisting of N turns of wire carrying a current I, shown in Fig. 7.4. By symmetry the field on the axis of the coil points along this axis. If we take the axis of this coil to be the z-axis then the field $B_z(z)$ on the z-axis at a distance z from the plane of the coil may be calculated using the Biot–Savart law to be

$$B_z(z) = \frac{1}{2} \frac{\mu_0 N I R^2}{(R^2 + z^2)^{3/2}}. \qquad [7.8]$$

The magnetic field at the centre of the coil and pointing along the z-axis is obtained by setting $z = 0$ in equation [7.8] to obtain

$$B_z(z = 0) = \frac{1}{2} \frac{\mu_0 N I}{R}. \qquad [7.9]$$

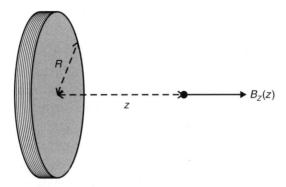

Fig. 7.4   A short plane circular coil of radius R has N turns of wire carrying a circulating current I. The magnetic field on the axis of the coil and a distance z from the plane of the coil is shown as $B_z(z)$ and is directed along the axis of the coil.

A particularly useful configuration for the generation of homogeneous fields is the **Helmholtz pair**. This consists of two identical plane circular coils of radius R consisting of N turns of wire wound in the same sense and carrying a current I. The planes of the coils are parallel and they share the same axis perpendicular to their planes which we will take to be the z-axis, as shown in Fig. 7.5. If the two coils are separated by a distance equal to R then the field on the z-axis and half-way between them (at 0 in the figure) is particularly homogeneous. For any separation of the two coils we would expect the total value of dB_z/dz, to be zero at that position because the two contributions of dB_z/dz, to this field gradient from the two coils will be of equal magnitude and of opposite sign. However, if the separation of the coils is fixed at R, the radius of the coils, then it can be

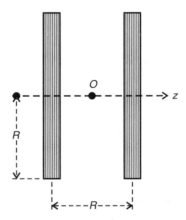

Fig. 7.5   A pair of plane circular coils which share the same axis in the Helmholtz configuration. The coils have a radius R, they each have N turns of wire carrying a current I circulating in the same sense and they are situated in parallel planes a distance R apart. The magnetic field near O is particularly homogeneous.

shown that the second derivative of the total axial field d^2B_z/dz^2, is also zero, which leads to a particularly homogeneous field in that region. The field is in fact not as homogeneous as that within a long solenoid, but the experimental advantage of the Helmholtz pair is that it allows free access from the side to a sample placed axially between the coils, which is not possible with the long solenoid.

7.5   Practical coils for the generation of laboratory magnetic fields

For a single circular coil of radius R, the variation in the magnetic field is less than 1% for a centrally placed cylinder with the same axis as the coil, a radius of $0.11R$ and a total length of $0.16R$. The field variation is less than 5% in a centrally placed cylinder of radius $0.25R$ and a total length of $0.36R$. These calculations were performed for a coil which consisted of ten layers of ten turns each, closely wound with wire of radius $0.001R$ so that the cross-section of the winding was $0.01R \times 0.01R$. The former on which the coils were wound had a radius $R = 1$. Provided that the dimensions of the cross-section of the coil are small in comparison with the radius, the shape of the coil cross-section has little effect. Thus for a small specimen, placing it at the centre of a single coil may well provide adequate homogeneity of the field.

For larger specimens where side-access to the specimen is required, a pair of Helmholtz coils may be adequate. For a pair of Helmholtz coils of mean radius R and separated along their common axis by an average distance R, the magnetic field varies by less than 1% in a centrally placed cylindrical region with the same axis as the coils which has a radius of $0.25R$ and a total length of $0.6R$. The field amplitude variation is less than 5% within a centrally placed cylinder of radius $0.4R$ and total length $0.9R$. These calculations were performed for two coils with the dimensions of the single coil described above with an average radius for the windings of $R = 1$ and a cross-section of $0.01R \times 0.01R$ for the windings. Again the results hold good provided that the dimensions of the cross-section of the windings are small in comparison with the radius of the coils.

For the highest field homogeneity, a solenoid should be used. For a solenoid with a length L large in comparison with its radius R, the homogeneity is better than 1% out to a radius of $0.95R$ over 80% of the length. However, a solenoid with $L \gg R$ may not provide adequate access to the specimen and good homogeneity may be obtained with solenoids with a length only a few times the diameter. As an example, a closely wound solenoid of radius R and length $5R$ produces a magnetic field with an amplitude variation of less than 1% in a centrally placed cylindrical region of radius $0.85R$ and a length of $0.22L$. The field amplitude variation is less than 5% over a centrally placed cylinder of radius $0.95R$ and total length $0.46L$. This calculation was performed for a solenoid closely wound in a single layer with wire of diameter $0.001R$. The field in the central region of this coil was 92.8% of the value $B = \mu_0 NI$, predicted for the long solenoid, which is almost exactly the value predicted by the theoretical expression for the magnetic field on the axis of a **finite** solenoid of these dimensions quoted in Problem 7.7. The homogeneity of the field within a short solenoid may be improved markedly by winding extra coils on the same cylindrical former just beyond the ends of the solenoid winding, to boost the field at the solenoid ends.

Topics covered in Chapter 7

1.   For simple current circuit geometries, the magnetic fields created can be deduced using only Ampère's and Gauss's laws.

2.   The magnetic field of a straight current-carrying wire forms closed circles, centred on the wire and in a plane perpendicular to it. The amplitude of the magnetic field circulating about the wire is proportional to the current in the wire and inversely proportional to the radius of the circle.

3.   A long solenoid, formed by closely winding wire on a former in the shape of a circular cylinder, creates a magnetic field everywhere within the cylinder which is uniform, directed along the axis of the cylinder and proportional in amplitude to the current flowing in the cylindrical coil and the number of turns of wire per unit length of the cylinder. Outside the cylinder the magnetic field moves in circles around the cylinder axis and has the same amplitude and direction as the field due to a single straight wire along the cylinder axis and carrying the same current as flows in the wire wrapped around the former. These calculations are valid only in regions remote from the ends of the solenoid.

4.   For more complicated circuit geometries the Biot–Savart law defines the magnetic field at a given position due to a short length of current-carrying wire oriented in a particular direction. The total magnetic field at that position created by the current-carrying wire can then be obtained by summing the effects of all the short sections that compose the total length of the wire using a line integral.

5.   Using the Biot–Savart law, the field created by a plane circular coil of many turns may be calculated at any position relative to the coil.

6.   A particularly homogeneous region of magnetic field may be generated by two circular coils located in planes perpendicular to the common axis of the two coils. If the two coils are wound in the same sense with the same number of turns and carry the same current and are located a distance apart along their common axis equal to their radius, they are called Helmholtz coils. The magnetic field near the axis of this configuration and centrally placed between the coils is particularly homogeneous.

7.   The coil configurations that create the largest volume of homogeneous field relative to the coil size are, in descending order, the long solenoid, the Helmholtz pair and the single plane coil. Although the Helmholtz pair creates magnetic fields that are not so homogeneous as those produced by the long solenoid, it does provide access to a specimen placed between the coils along the system axis and in all directions perpendicular to this axis which is not possible with the long solenoid.

Important equations of Chapter 7

1.   The magnetic field $\tilde{B}(R)$ at a distance R from a long straight wire carrying a current I forms circles centred on the wire which are located in planes perpendicular to the direction of the wire. The amplitude of $B(R)$ is given by

$$B(R) = \frac{\mu_0 I}{2\pi R}.$$

2.   The magnetic field $\tilde{B}$ anywhere inside a long cylindrical solenoid, closely wound with wire carrying a current I with N turns of wire per unit length of the solenoid, has a constant magnitude given by

$$B = \mu_0 N I.$$

The field is directed along the axis of the solenoid in the direction of advance of a right-hand thread rotated in the direction that the current circulates. Outside the solenoid the field circulates in plane circles about the axis of the solenoid and has the same amplitude as the field that would be generated by a current I flowing along the axis of the solenoid. These results only hold in regions remote from the ends of the solenoid.

3.   The Biot–Savart law yields the magnitude and direction of the magnetic field increment $dB(R, \theta)$ created by a current $\tilde{I}$ flowing in a short section of a conductor of length dL. At the field point $P(R, \theta)$, defined by the polar co-ordinates shown in Fig. 7.3a with the origin of the polar coordinates located in dL, the field increment is given by

$$dB(R, \theta) = \frac{\mu_0 I dL \sin(\theta)}{4\pi R^2} = \frac{\mu_0 dL}{4\pi R^3} |[\tilde{I} \times \tilde{R}]|.$$

The angle θ is that between the directions of the short length of conductor dL and R, the distance from dL to the field point $P(R, \theta)$. The field is perpendicular to the directions of both $\tilde{I}$ and $\tilde{R}$ and points in the direction of a right-hand thread rotated from the direction of $\tilde{I}$ or dL to that of $\tilde{R}$ as signalled by the presence of the vector product, which is described in Appendix 1. To calculate the total magnetic field at the field point the effects of all the elemental current-carrying lengths dL must be integrated over the whole length of the current-carrying conductor.

4.   The magnitude of the axially directed magnetic field, on the axis of a plane circular coil of wire of radius R, which consists of N turns of wire carrying a current I, at a point a distance z from the plane of the coil, is

$$B(z) = \frac{1}{2} \frac{\mu_0 N I R^2}{(R^2 + z^2)^{3/2}} \, .$$

5.   Two identical plane circular coils wound in the same sense and carrying the same current, which share the same axis, produce a particularly homogeneous field on their common axis half-way between them, if they are separated by a distance equal to their radius. This configuration is called the Helmholtz pair and is illustrated in Fig. 7.5.

Problems

7.1   A straight conductor with a circular cross-section of radius a carries a total current I, which may be assumed to be distributed uniformly over the cross-section of the conductor. Using Ampère's and Gauss's laws, deduce the direction and strength of the magnetic field inside the conductor at a distance R ($R < a$) from its axis. Using the result you obtain and equation [7.1], sketch a graph of the magnetic field amplitude against R both inside and outside the conductor.

7.2   Two long straight thin wires, each carrying a current I, lie in the (x, y) plane. One lies along the line $y = a$ and has the current directed parallel to the $+x$-axis and the other lies along the line $y = -a$ and has the current directed anti-parallel to the $+x$-axis. Find the magnitude and direction of the magnetic field at positions in the (x, y) plane with coordinates $(0, 0)$, $(0, a/2)$ and $(0, 2a)$. At a position $(0, y_1)$ how would you expect the amplitude of the field to vary as y_1 becomes large in comparison with a?

7.3   A coaxial cable consists of a thin central wire and a surrounding thin conducting circular cylinder of radius a with the wire as axis. The signal current flows I along the central wire and the same total current returns through the surrounding cylinder. Show that this arrangement generates no magnetic field external to the cylinder. What are the direction and amplitude of the magnetic field within the cylinder at a distance R ($R < a$) from the wire?

7.4   Two straight thin cylindrical conductors are parallel and their axes are separated by a distance d. They each carry a current I. Calculate the force per unit length exerted by one wire on the other. What is the direction of the force if the current flows are (a) parallel and (b) anti-parallel?

7.5   Verify directly, using the Biot–Savart law, that the magnetic field at the centre of a plane circular coil of radius R, which has N turns of wire each carrying a current of I, is directed axially and has an amplitude given by equation [7.9].

7.6   A plane circular coil is located in a vertical plane that includes the direction of the Earth's magnetic field. At its centre is located a magnetic compass needle freely pivoted to rotate in a horizontal plane about a vertical axis. The coil has a diameter of 0.1 m and has 10 turns of wire each carrying a current of 0.25 A. When the current flows through the coil the compass needle rotates by $45°$. What is the local magnitude of the horizontal component of the Earth's magnetic field?

7.7   A solenoid wound on a cylindrical former has N turns of wire per unit length, carrying a current I. The axis of the solenoid is the z-axis. It has a length L so that it reaches from $z = 0$ to $z = L$. At any point with coordinates $(0, 0, z)$ on the z-axis, the upper and lower circular plane ends of the solenoid subtend circular cones of semi-angles ϕ_1 and ϕ_2, respectively. It can then be shown that the z-directed field B_z at the point $(0, 0, z)$ such that $0 \leq z \leq L$ is given by the expression

$$B_z = \frac{1}{2} \mu_0 I N [\cos(\phi_1) + \cos(\phi_2)].$$

Verify that this formula gives the correct values for the axially directed field (a) on the axis of an infinitely long solenoid and (b) on the axis and at the end of a semi-infinite solenoid. This formula is useful in giving the field on the axis within solenoids of finite length.

7.8   A pair of Helmholtz coils have a radius R and each consists of N turns of wire carrying a current I. Show that the magnetic field on the common axis of the coils and half-way between them is given by B in the expression

$$B = 0.7155 \frac{\mu_0 N I}{R}.$$

8. Magnetic polarization of material

In the same manner that electrically polarizable material complicated the application of Gauss's law in electrostatics, the presence of magnetically polarizable material complicates the application of Ampère's law in magnetostatics. However, as with electrostatics, the introduction of a new field vector enables this complication to be simply overcome in magnetism. The properties of the new field vector are discussed and contrasted with those of the magnetic field. The continuity of the new vector and the magnetic field vector across the boundary of two adjacent blocks of different magnetically polarizable material is described.

8.1 Magnetic material

Every solid, liquid or gas is **electrically polarizable** because the electron clouds around each atom or molecule may be distorted in an electric field. Some molecules, such as water, have, in addition, permanent electric dipoles that may orient in an electric field, as we discussed in Chapter 4. The same is not true in magnetism where very few materials are **magnetically polarizable** to an appreciable extent by an applied magnetic field. As a result, magnetic fields penetrate most material very little changed and Ampère's law holds to good accuracy in the form we have already derived for magnetic fields in a vacuum in equation [6.19]. For these reasons students who are primarily interested in biological effects of electric and magnetic fields could skip this chapter and pass on to Chapter 9, which has direct relevance to biology, without losing much relevant understanding.

The reason for the qualifications in the above statements that most material is not magnetically polarizable to an 'appreciable extent' or that Ampère's law holds to 'good accuracy' is that there does exist a mechanism for the magnetic polarizability of most substances, but the effects are very weak. The effect is called **diamagnetism** and it has its origin in the fact that an applied magnetic field does, to a very small extent, alter the motion of an electron circulating about its nucleus and induces a very small magnetic moment in the atom directed oppositely to the inducing magnetic field. We will deal with this effect in Chapter 10, but for our present purposes we need only note that for light elements in all but the most powerful artificially produced magnetic fields, the induced diamagnetic magnetic moment has a negligible effect.

The angular momentum and spin of electrons in atomic orbits may give rise to an atomic magnetic moment in some atoms. Semi-classically, the orbiting or spinning electron may be thought of as forming a microscopic current loop. However, the inner filled electron shells of most atoms contain pairs of electrons with equal and opposite angular momentum and no total magnetic moment results. The outer 'valence' electrons are shared with neighbouring atoms in fixed orbits to form chemical bonds, and these too give rise to no magnetic moment. Permanent atomic magnetic moments are found among the elements only in a few places in the periodic table where electrons are added to outer electron shells leaving incomplete inner electron shells. The most important group is the 'iron group' of elements which includes iron, cobalt and nickel, within which the 3d electron shell is incomplete. Solid, liquid or gaseous material containing such elements may contain elementary magnetic moments that can be aligned in an applied magnetic field and so give rise to a field-induced magnetic moment. Such material is called **paramagnetic**, and a field-induced magnetic moment per unit volume comes about by the partial alignment of the permanent magnetic dipole moments of the atoms which is opposed by the disturbing effects of the thermal fluctuations. We discussed in Chapter 5 a very similar situation for the electric dipole moment per unit volume induced by an electric field acting on the permanent electric dipole moments of the water molecules. In most paramagnetic material at room temperature the paramagnetic moment induced by a given magnetic field is much bigger than the diamagnetic moment induced in the same material by the same field. In this chapter we will discuss for completeness how to deal with magnetically polarizable material, even if this effect is seldom of much importance in biology.

A third class of magnetic material consists of **ferromagnetic, anti-ferromagnetic and ferrimagnetic materials**, and these may possess very powerful permanent magnetic moments even in the absence of an applied magnetic field. The effect is due to a quantum mechanical co-operative coupling between the spins of the atoms in a crystal, and such material may not be dealt with using the techniques to be developed in this chapter. One such magnetic material that does play an important role in biology is **magnetite**, which is synthesized by many living systems and which we will discuss as a possible animal magnetic compass in Chapter 14.

8.2   Modification of Ampère's law by induced magnetic moments

Just as the presence of dielectrics complicated our original formulation of Gauss's law in electrostatics, the presence of field-induced magnetic moments complicates the application of Ampère's law in magnetostatics. An induced magnetic moment is formally equivalent to an induced current loop, as expressed in equation [6.3], and

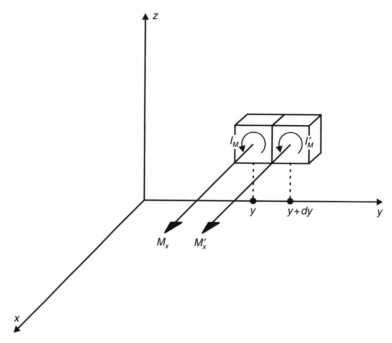

Fig. 8.1 Two adjacent small blocks of magnetically polarizable material with volumes $dxdydz$ such that the y-coordinates of their centres are y and $y + dy$ respectively. The magnetic moments per unit volume induced at the centres of the blocks by a magnetic field with a component along the x-axis are directed along the x-axis and have amplitudes M_x and M'_x respectively. The magnetizations of the blocks may be considered as due to magnetization currents I_M and I'_M circulating around them as shown.

we must take account of such induced currents in our formulation of Ampère's law. Let the induced magnetic moment per unit volume at the position (x, y, z) be represented by the vector $\tilde{M}$. If we consider only M_x, the x-component of this vector as depicted in Fig. 8.1, the total x-component of the magnetization in the small volume $dxdydz$ at (x, y, z) may be written as $M_x dxdydz$. This magnetic moment may be thought of as due to a magnetization current I_M, circulating anti-clockwise, as we view it, about this small volume such that the current I_M multiplied by the area of the loop $dydz$ generates the induced magnetic moment $M_x dxdydz$ along the x-axis of the small volume. Thus we have in Fig. 8.1

$$I_M dydz = M_x dxdydz \quad \text{or} \quad I_M = M_x dx. \qquad [8.1]$$

Let us now turn to the neighbouring small volume, shown in Fig. 8.1, which is displaced by dy along the y-axis where the induced magnetization in the x-direction may be written as $M_x + (dM_x/dy)dy$. Here dM_x/dy is simply the rate of change of

M_x with y. Repeating the argument that led to equation [8.1], for this second volume we obtain an expression for the current I'_M circulating anti-clockwise about this volume as

$$I'_M = \left[M_x + \left(\frac{dM_x}{dy} \right) dy \right] dx.$$ [8.2]

There is thus partial cancellation of the currents that flow along the face shared by these two small volumes, so that the total current flowing in that face is given by

$$I_M - I'_M = -\left(\frac{dM_x}{dy} \right) dxdy.$$ [8.3]

As can be seen in Fig. 8.1, the current $I_M - I'_M$ flows in the positive z-direction. If we represent this z-directed current by a current density, or current per unit area, J_{mz}, flowing through the area $dxdy$ perpendicular to the z-axis, we have the relation that $I_M - I'_M = J_{mz}dxdy$ which, combined with equation [8.3], yields the relation

$$J_{mz} = -\frac{dM_x}{dy}.$$ [8.4]

We would obtain another contribution to the z-directed current density J_{mz}, namely $+dM_y/dx$, if we were to consider in an identical fashion the variation in the magnetization M_y induced in the y-direction in two adjacent small volumes $dxdydz$ displaced from each other by a distance dx along the x-axis. The total value of J_{mz} then becomes

$$J_{mz} = \left(\frac{dM_y}{dx} - \frac{dM_x}{dy} \right).$$ [8.5]

In a similar fashion we may obtain expressions for the other components J_{mx} and J_{my} of the total current density, directed along the x and y coordinate axes, that are equivalent in effect to the total magnetic moment $\tilde{M}$ induced in a general direction as

$$J_{mx} = \left(\frac{dM_z}{dy} - \frac{dM_y}{dz} \right) \quad \text{and} \quad J_{my} = \left(\frac{dM_x}{dz} - \frac{dM_z}{dx} \right).$$ [8.6]

A vector which has the x-, y-, and z-components shown on the right-hand side of equations [8.6] and [8.5] may be written as curl($\tilde{M}$) in vector notation (see Appendix 1). As with div($\tilde{P}$), we will not assume any knowledge of vector algebra but will simply use the expression curl($\tilde{M}$) as a short way of writing a vector which has a z-component as in equations [8.5] and x- and y-components as in equation [8.6]. We may then write equations [8.5] and [8.6] with their components written in full, or in compact notation, as

$$\tilde{J}_M = [J_{mx}, J_{my}, J_{mz}]$$

$$= \left[\left(\frac{dM_z}{dy} - \frac{dM_y}{dz} \right), \left(\frac{dM_x}{dz} - \frac{dM_z}{dx} \right), \left(\frac{dM_y}{dx} - \frac{dM_x}{dy} \right) \right] = \text{curl}(\tilde{M}). \quad [8.7]$$

Equation [8.7] gives the direction and strength of the induced current density $\tilde{J}_M$ at a given location that will reproduce the magnetization per unit volume $\tilde{M}$ that is induced at that location.

In electrostatics we invoked in Chapter 4 the purely geometrical divergence theorem to modify Gauss's law so as to include the effects of induced electric dipole moments. Here, in a similar manner, we will invoke the purely geometrical **Stokes's theorem** to modify Ampère's law so as to include the effects of induced magnetic dipole moments. Stokes's theorem states that the line integral of any vector $\tilde{v}$ along a given closed path is equal to the total flux of $\text{curl}(\tilde{v})$ through any surface that is bounded by that closed path (see Appendix 1). We may write this as

$$\oint v_R dR = \int \int_S [\text{curl}(\tilde{v})]_N dS. \quad [8.8]$$

Let us now consider a situation where there exist applied magnetic fields in the presence of magnetically polarizable material. We suppose that the applied magnetic fields are generated by currents $\tilde{I}$ that we specify in strength and position. If we apply Ampère's law to this region of space we must clearly take account of the original magnetic fields generated by the currents $\tilde{I}$, but we must also take account of the additional magnetic fields that are generated by the magnetic moments per unit volume $\tilde{M}$ induced in the magnetically polarizable material by the original applied fields. We can represent the effects of the induced magnetic moments $\tilde{M}$ by including in the statement of Ampère's law the equivalent induced current densities $\tilde{J}_M$ that we deduced in equation [8.7] as well as the current densities $\tilde{J}$ that represent the original applied currents $\tilde{I}$. When we do so Ampère's law, as in equation [6.19], becomes

$$\oint B_R dR = \mu_0 \int \int_S (\tilde{J} + \tilde{J}_M)_N dS. \quad [8.9]$$

In analogy with the electrical case, we will describe the magnetization current density $\tilde{J}_M$ as **bound current density**, as it is fixed where it was induced, and the original external magnetic-field-creating current density $\tilde{J}$ as **free current density**, as we are free to locate this where we wish.

From equation [8.7] we may substitute $\tilde{J}_M = \text{curl}(\tilde{M})$ so that equation [8.9] becomes

$$\oint \frac{B_R}{\mu_0} dR = \int\int_S [\tilde{J} + \tilde{J}_M]_N dS = \int\int_S [\tilde{J} + \text{curl}(\tilde{M})]_N dS. \qquad [8.10]$$

Using Stokes's theorem in the form of equation [8.8], we may transform the flux of $\text{curl}(\tilde{M})$ through the surface S into the line integral of $\tilde{M}$ around the path that defines the surface boundary so that we have

$$\int\int_S [\text{curl}(\tilde{M})]_N dS = \oint M_R dR.$$

If we now take this term to the left-hand side of the equation, we obtain

$$\oint \left(\frac{\tilde{B}}{\mu_0} - \tilde{M} \right)_R dR = \int\int_S J_N dS. \qquad [8.11]$$

If we now define a new vector

$$\tilde{H} = \left[\frac{\tilde{B}}{\mu_0} - \tilde{M} \right], \qquad [8.12]$$

then we may write Ampère's law simply as

$$\oint H_R dR = \int\int_S J_N dS. \qquad [8.13]$$

   This is a more general form of Ampère's law in magnetostatics than [6.19] because it holds whether there exists magnetically polarizable material or not. In words, it says that **the line integral of the vector $\tilde{H}$ around any closed path is equal to the total flux of the free current passing through any surface bounded by that path**. The relationship between the magnetic field $\tilde{B}$ and the new vector $\tilde{H}$, defined in equation [8.12], may be written as

$$\tilde{B} = \mu_0(\tilde{H} + \tilde{M}). \qquad [8.14]$$

In homogeneous material where the induced magnetization $\tilde{M}$ is in the direction of the inducing magnetic field $\tilde{B}$, all three vectors $\tilde{B}, \tilde{M}$ and $\tilde{H}$ are parallel and we may write equation [8.14] as

$$\tilde{B} = \mu_R \mu_0 \tilde{H}, \qquad [8.15]$$

where μ_R is a scalar constant, characteristic of the magnetically polarizable material at a particular location, and is called the **relative permeability**. It takes the place of the relative dielectric constant in electricity. As mentioned in Section 8.1, for the vast majority of biological materials the induced magnetization M is close to zero and thus μ_R is close to 1. In these circumstances Ampère's law in the form of equation [6.19] remains applicable and there is no purpose in introducing the new vector $\tilde{H}$.

8.3   Properties of the vector $\tilde{H}$

Although some of the mathematical manipulations in this chapter are complicated, the logic is simple and parallels exactly that in the modification of Gauss's law in electricity to take account of the presence of dielectrics. In the presence of field-induced magnetization, the application of Ampère's law is complicated by induced magnetization $\tilde{M}$ or, exactly equivalent, bound currents $\tilde{J}_M$. As in electricity the induced magnetic moments will create extra magnetic fields and it may at first be thought that some iterative procedure will be necessary. However, by invoking Stokes's theorem and by the invention of a new field variable $\tilde{H}$, we are able to rewrite Ampère's law in a simple way that takes full account of the induced magnetization. The price paid for the increased applicability of the law is the introduction of the new field $\tilde{H}$. As with the invention in electricity of the new vector $\tilde{D}$ to replace $\tilde{E}$ in Gauss's law, it is important to distinguish clearly between the magnetic field $\tilde{B}$ and the new vector $\tilde{H}$. Again as in electricity, the new vector field $\tilde{H}$ is non-physical in that it is created by free currents and not by bound currents, whereas both types of current correspond to charge circulation and are equally real physically. Also, like electricity, where the force on a charge is determined by $\tilde{E}$ and not $\tilde{D}$, the force on a current or a magnetic dipole is always determined by $\tilde{B}$ and never by $\tilde{H}$. The force vectors that express the real physical effects of electric and magnetic fields are thus always $\tilde{E}$ and $\tilde{B}$.

A recipe for studying the effects of magnetic fields in the presence of paramagnetically permeable material is as follows:

1. Use Ampère's law as in equation [8.13] to deduce the value of $\tilde{H}$ due to the free currents that define the magnetic situation.
2. Use the relationship between $\tilde{H}$ and $\tilde{B}$ as in equation [8.15] and the known value of the local relative permeability μ_R to find the value of the real physical vector $\tilde{B}$.
3. To find the values of $\tilde{B}$ and $\tilde{H}$ in contiguous material with a different relative permeability, use the boundary conditions that govern how the vectors $\tilde{B}$ and $\tilde{H}$ change between material of different relative permeabilities which we will derive in the next section.

8.4   Boundary conditions for $\tilde{B}$ and $\tilde{H}$

In Fig. 8.2a we show the plane boundary between materials 1 and 2 which have relative permeabilities of μ_{R1} and μ_{R2}, respectively. We construct a Gauss law surface in the form of a circular cylinder with its axis perpendicular to the interface and with its flat ends buried within the two materials. Also shown in the diagram are the components of the magnetic field perpendicular to the interface B_{N1} and B_{N2} in the two materials. By invoking Gauss's law we may deduce that the total flux of $\tilde{B}$ out of the cylinder is zero, so that

$$B_{N1} = B_{N2}. \tag{8.16}$$

In this derivation we have used the fact that we can make the cylinder as short as we like to exclude any flux of $\tilde{B}$ through the cylinder walls due to a component of $\tilde{B}$ that is not normal to the interface.

In Fig. 8.2b we again show the interface between the same two materials but now define a rectangular path *abcda* with sides *ab* and *cd* parallel to the interface and sides *bc* and *da* very short and perpendicular to the interface. The tangential component of $\tilde{H}$ in material 1 is represented by H_{T1} and the parallel component of $\tilde{H}$ in material 2 is represented by H_{T2}. Provided there exists no **free** current flowing along the interface we can use Ampère's law as in equation [8.13] to deduce that the line integral of H around this circuit is zero. We can make any contribution from sections of the path

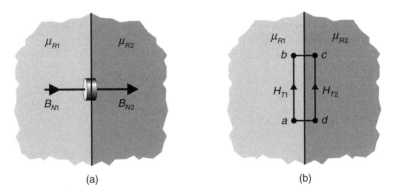

(a)   (b)

Fig. 8.2   (a) The plane interface between materials with relative permeabilities of μ_{R1} and μ_{R2}. A cylindrical Gauss law surface is shown buried in the interface. By invoking Gauss's law in the magnetic field, it is shown the normal components of the vector $\tilde{B}$ are conserved across the interface so that $B_{N1} = B_{N2}$. (b) A rectangular circuit *abcda* is shown that spans the plane interface between materials with different relative permeabilities. Provided that no free current flows along the interface, Ampère's law decrees that $\oint H_R dR = 0$, from which it is shown that the tangential components of the vector $\tilde{H}$ are conserved across the interface so that $H_{T1} = H_{T2}$.

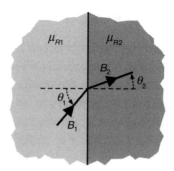

Fig. 8.3   The diffraction of the magnetic field at the plane interface between materials with different relative permeabilities. A magnetic field B_1 which approach the interface by making an angle θ_1 to the normal leaves the interface as B_2 which makes an angle θ_2 with the normal.

bc and da to the line integral negligible by reducing the length of these sections. Thus we may deduce that

$$ab \cdot H_{T1} - cd \cdot H_{T2} = 0$$

$$\text{or} \quad H_{T1} = H_{T2} \quad \text{as} \quad ab = cd. \qquad [8.17]$$

Thus, in words, the boundary conditions on $\tilde{B}$ and $\tilde{H}$ as we move between two materials of different relative permeability may be stated simply as **normal $\tilde{B}$ is conserved as is tangential $\tilde{H}$**.

Just as in the electrical case (see equations [5.6] to [5.8]), it is easy to show, using these boundary conditions, that the magnetic field $\tilde{B}$ is diffracted as it passes through the plane interface between two materials with different relative permeability. If in material 1 the field makes an angle θ_1 with the normal to the surface it will make an angle θ_2 with the normal in material 2 such that

$$\tan(\theta_1) = \left[\frac{\mu_{R1}}{\mu_{R2}}\right] \tan(\theta_2). \qquad [8.18]$$

This situation is shown in Fig. 8.3.

Topics covered in Chapter 8

1.   The magnetic moment per unit volume induced by an external magnetic field in most materials, and nearly all biological material, is very small. When dealing with such material, Ampère's law, as derived in Chapter 6 for a vacuum, still holds with good accuracy.

2.   A few materials contain atoms with only partially filled inner electron shells, and associated with such atoms are appreciable induced magnetic moments in an external

magnetic field. These materials are called paramagnetic. The magnetic moments are caused by the partial alignment of permanent atomic magnetic moments in the applied magnetic field which is disturbed by random thermal motions.

3.   When a paramagnetic substance is placed in an externally generated magnetic field, a magnetic dipole moment per unit volume is created within it. These induced magnetic dipole moments generate additional magnetic fields which must be taken into account when applying Ampère's law.

4.   In the presence of magnetically polarizable material Ampère's law may be expressed simply in terms of a new magnetic field vector $\tilde{H}$. The law then states that the line integral of the vector $\tilde{H}$ taken around any closed path is equal to the flux of free current that passes through any surface that is bounded by the chosen path.

5.   The new vector $\tilde{H}$ is created by free currents but not by bound currents and is thus non-physical because both types of currents are equally real physically.

6.   To predict the outcome of magnetic experiments performed in the presence of magnetically polarizable material the following steps are often necessary. First, the amplitude and direction of the vector $\tilde{H}$ are deduced using Ampère's law and a knowledge of the free currents that define the magnetic set-up. Then the amplitude and direction of the magnetic field $\tilde{B}$ are deduced from the relationships that exist between the amplitudes of B and H mentioned below. Finally, the physical forces and energies involved may be deduced from the knowledge of the amplitude and direction of the real magnetic field $\tilde{B}$.

7.   Within homogeneous and isotropic solids and in liquids and gases, the magnetic field $\tilde{B}$ and the new field vector $\tilde{H}$ point in the same direction, and the amplitude of the magnetic field B is obtained from the amplitude of H simply by multiplication by μ_0 and by a numerical constant μ_R characteristic of the material.

8.   At the plane boundary of two materials with different magnetic polarizabilities, the normal component of the magnetic field $\tilde{B}$ and the tangential component of the field $\tilde{H}$ are conserved. As a result a magnetic field that is directed towards the plane boundary obliquely, changes direction abruptly at the boundary.

Important equations of Chapter 8

1.   A magnetic field $\tilde{B}$ will induce a magnetic moment $\tilde{M}$ per unit volume in a magnetically permeable material. The induced magnetization at a given location may be represented by a bound current density $\tilde{J}_M$ at that location which is related to $\tilde{M}$ through the relation

$$\tilde{J}_M = \mathrm{curl}(\tilde{M}).$$

The components of the vector curl($\tilde{M}$) are given in equations [8.5] and [8.6] and Appendix 1.

2.   In the presence of magnetically permeable material, Ampère's law states that the line integral of the vector $\tilde{H}$ around any path is equal to the total flux of **free current** density $\tilde{J}$ cutting any surface bounded by that path:

$$\oint H_R dR = \int\int_S J_N dS.$$

3.   The vector $\tilde{H}$ is related to the magnetic field $\tilde{B}$ by the relation

$$\tilde{B} = \mu_0(\tilde{H} + \tilde{M}) \quad \text{or} \quad \tilde{B} = \mu_R \mu_0 \tilde{H},$$

where μ_R is the relative permeability of the material at that point. The second expression holds only for isotropic solid material and for liquid and gaseous material within which $\tilde{B}$, $\tilde{H}$ and $\tilde{M}$ all point in the same direction.

4.   At the boundary between materials with different relative permeabilities μ_R, the boundary conditions that determine the changes in $\tilde{B}$ and $\tilde{H}$ across the boundary state that the normal component of $\tilde{B}$ and the tangential component of $\tilde{H}$ are conserved. As a result, the direction of a magnetic field $\tilde{B}$, which is not normal to the interfacial surface, changes direction abruptly as it passes through such a boundary.

Problems

8.1   The (x, y) plane separates two blocks of material. For $z > 0$ the relative permeability is μ_{R1} and for $z < 0$ the relative permeability is μ_{R2}. A current flows along the interface in the y-direction such that the total current per unit width in the x-direction is J_y. Using Ampère's law, deduce how the values of H_x and H_y vary as the interface is crossed.

8.2   An applied magnetic field $\tilde{B}$ is directed normally to the plane face of a block of magnetic material with a relative permeability of μ_R. Show that the magnetic moment per unit volume M induced in the material is given by

$$M = \frac{B}{\mu_0}\left(1 - \frac{1}{\mu_R}\right).$$

8.3   Following an argument similar to that used in Problem 1.5, show that at any point the components of a static magnetic field must obey a relationship of the form

$$\text{div}(\tilde{B}) = \frac{dB_x}{dx} + \frac{dB_y}{dy} + \frac{dB_z}{dz} = 0.$$

This is the microscopic form of Gauss's law for magnetic fields.

8.4   When a small spherical specimen made of paramagnetic material is located at a position where the components of the magnetic field $\tilde{B}$ are (B_x, B_y, B_z), a magnetic moment $\tilde{m}$ with components (m_x, m_y, m_z) is induced in the specimen such that

$$m_x = KB_x, \quad m_y = KB_y, \quad m_z = KB_z.$$

K is a constant, characteristic of the material, which takes the value $K = \kappa v/\mu_0$. In this expression v is the volume of the specimen and κ is the magnetic susceptibility. If the force on the specimen in the x-direction is given by F_x in the expression

$$F_x = m_x \frac{dB_x}{dx} + m_y \frac{dB_y}{dx} + m_z \frac{dB_z}{dx},$$

show that F_x may be written

$$F_x = \left(\frac{\kappa v}{2\mu_0} \right) \frac{dB^2}{dx}.$$

Hence deduce whether paramagnetic material, for which $\kappa > 0$, is attracted toward locations where the magnetic field is large or repelled from them.

8.5*   A simple form of an electromagnet is shown in Fig. 8.4. It consists of a soft iron rod with a uniform circular cross-section of radius a and with a relative permeability μ_R, which is bent into a circle such that the axis of the rod forms a circle of radius R. A parallel-sided gap of small width d is left between the ends of the rod as shown in

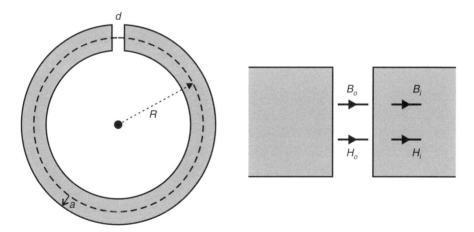

Fig. 8.4   On the left is shown the core of an electromagnet which consists of a cylindrical iron rod of radius a bent into a circle of radius R but leaving a parallel faced gap of width d. On the right is shown an enlarged view of the gap in the magnet core. In the gap the magnetic field vectors are B_o and H_o and in the iron they are B_i and H_i.

the figure. A coil of wire of N turns and carrying a current I is wound on the rod. Estimate the magnetic field B_o in the gap, assuming that $a \ll R$ and $d \ll R$.

Worked solution

Let the magnetic fields in the gap be represented by B_o and H_o and those within the iron bar be represented by B_i and H_i, as sketched in the enlarged view of the gap shown in Fig. 8.4. Normal $\tilde{B}$ is conserved so that, using the relationship $\tilde{B} = \mu_R \mu_0 \tilde{H}$ we may deduce that

$$B_o = B_i \quad \text{and} \quad H_o = \mu_R H_i.$$

If we apply Ampère's law (see equation [6.19]) to the circular path of radius R, shown as a broken line in the figure, we obtain the result

$$(2\pi R - d)H_i + dH_o = NI$$

$$\text{or} \quad H_o = \frac{NI}{\frac{(2\pi R - d)}{\mu_R} + d}.$$

From this equation we find the required magnetic field in the gap to be

$$B_o = \mu_0 H_o = \frac{\mu_0 NI}{\frac{(2\pi R - d)}{\mu_R} + d} \simeq \frac{\mu_0 \mu_R NI}{2\pi R + \mu_R d}.$$

To obtain this result I have assumed that B_i and H_i are approximately constant within the iron. Can you justify this assumption?

9. Induced electric and magnetic fields

In this chapter it is shown that a time-varying magnetic field will give rise to a time-varying electric field and a time-varying electric field will give rise to a time-varying magnetic field. Thus when the fields vary with time it is no longer possible to separate electricity and magnetism as was possible for static fields. Faraday's law, which governs the generation of an electric field by a time-varying magnetic field, will give rise to electric fields in the form of closed loops which can penetrate conductors, unlike the static electric fields dealt with in Chapter 2. Thus the generation of an electrical current within a living biological cell is often accomplished by applying an external time-varying magnetic field. The realization that a time-varying electric field will generate a magnetic field enabled James Clerk Maxwell to formulate a set of equations called the Maxwell equations that govern electric and magnetic fields under any circumstances.

9.1 Faraday's law of induction

Our treatments of static electric and magnetic fields have so far been separate. When we deal with fields that vary with time this clear separation is no longer possible. As we will show in this chapter, a time-varying magnetic field will give rise to an electric field and a time-varying electric field will give rise to a magnetic field. We will deal first with the generation of an electric field by a time-varying magnetic field.

In the first half of the 19th century it was noticed that moving a bar magnet in the vicinity of a coil of wire could result in the generation of a current in the coil. After much experimentation this effect was condensed by Michael Faraday into a single equation, called **Faraday's law**, which states that

$$\oint E_R dR = -\frac{d}{dt}\left[\int\int_S B_N dS\right]. \tag{9.1}$$

In words, it states that **the line integral of the electric field around any closed path is equal to minus the rate of change of the flux of the magnetic field $\tilde{B}$ through any surface bounded by that path**. Note that in order to observe the effect experimentally it is necessary to place a conducting wire to define the path, but the law holds around any path whether there is a conductor there or not. The line integral of the electric field around a given path is called the **electromotive force or emf**. It is expressed in volts and measures the work that would be done by the electric field on a unit charge passing once around that path. If we place a loop of conductor in a region where the magnetic field is varying with time, a current will be induced in the conducting loop. If the conducting loop has a resistance R the current I generated in this loop by the emf is given by

$$I = \frac{\text{emf}}{R}. \tag{9.2}$$

As seen by the definition of an electromotive force given above and by equation [9.2], an emf is similar to a voltage change. Both represent the work done by an electric field on a unit charge as it moves along a specified path. For electric fields created by static charges the field originates on positive charges and terminates on negative charges and the line integral of the field around any closed loop is zero, as shown in equation [1.12]. As we shall see, the electric fields created by time-varying magnetic fields take the form of closed loops and the line integral of the electric field around any chosen closed path is non-zero and given by equation [9.1]. If we place a closed loop of conductor in a region where the magnetic field varies with time we cannot specify the voltage at any particular point on the conductor, but we can say that the emf around the loop is given by equation [9.1] and that this emf will induce in the conductor a current given by equation [9.2]. The induced emf has a particular direction which is best defined by stating that the magnetic field created by current induced in a loop of conductor is in a direction such as to oppose the direction of the inducing magnetic field. This result is called **Lenz's law**.

Faraday's law is remarkable because this simple law predicts the emf around a given path whether the rate of change with time of the flux of $\tilde{B}$ threading the path is due to translation of the path, or to changing the amplitude of the inducing magnetic field at its source while keeping the path fixed in space. Notice also that Faraday's law gives the line integral of the electric field around a given path but does not define the strength of the electric field at any given point.

Unlike static electric fields which originate and terminate on charges, the electric fields induced by time-varying magnetic fields form closed loops. In this respect they resemble magnetic fields, which is no surprise because of the similarity in form between Faraday's law as in equation [9.1] and Ampère's law as written in equation [6.19]. Equation [9.1], which is universal in application, replaces the statement in equation [1.12] that the line integral of the electric field is zero around any closed path, which is only true in the absence of time-varying magnetic fields. Thus, in the presence of time-varying magnetic fields, when the line integral of the electric field around a closed loop is no longer zero, we can no longer assign a unique voltage to a given position (see equations [1.11] and [1.12] and the text between them). In these circumstances the electrostatic potential or voltage loses much of its value.

9.2   An application of Faraday's law

As an example of the application of Faraday's law, we will consider an apparatus often used to investigate the possible effects of the alternating magnetic fields generated by the domestic electricity supply on living cells. A circular plastic Petri dish filled with an aqueous solution is placed within a long solenoid such that the dish and solenoid are cylindrically symmetric about the same z-axis, as sketched in Fig. 9.1a. We will assume that the solenoid is fed with a current that varies sinusoidally with time so that it creates a sinusoidal alternating magnetic field. This magnetic field is directed along the z-axis, and we will further assume that it has a constant amplitude B_0 all over the dish. If the frequency of the alternating field is f cycles per second, we may represent the magnetic field as

$$B_z = B_0 \cos(2\pi ft).$$ [9.3]

As in Section 7.1 where we deduced the direction of the magnetic field near a straight current-carrying wire, this problem has cylindrical symmetry. The induced electric field at a distance R from the axis of the solenoid could have a radial

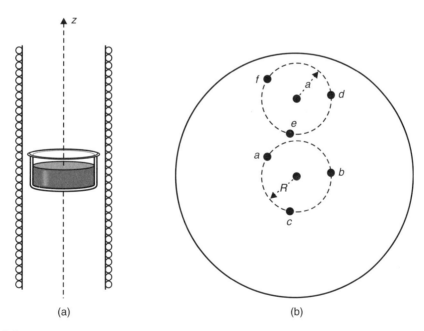

(a)   (b)

Fig. 9.1   (a) Typical experimental magnetic exposure system apparatus consisting of a vertical solenoid with a horizontal dish, co-axial with the solenoid, containing the material to be exposed to the magnetic field. (b) A plan view of the exposure dish with two circular circuits, *abca* and *defd*, shown in the plane of the dish.

component E_R, a tangential component E_T moving in circles around the wire or a component E_z parallel to the axis of the solenoid. By constructing a cylindrical Gauss law surface with the same axis as the solenoid and applying Gauss's law for an electric field to the uncharged fluid contained within that surface, we can eliminate any radial component of the induced electric field E_R. The outward fluxes of the electric field through the two plane ends of the Gauss law cylinder sum to zero because one is inward and the other outward and the applied field does not depend upon the coordinate z. Similarly, we can eliminate any axial or z-component of the induced electric field E_z by constructing a rectangular path in a plane that contains the axis of the solenoid with two sides parallel to the axis of the dish and two connecting radial sides and invoking the fact that the line integral of $\tilde{E}$ must be zero around this path as it encloses no changing flux of the applied magnetic field. To deduce the amplitude of the tangentially induced electric field E_T which rotates in circular paths centred on the solenoid axis and occupying planes perpendicular to the magnetic field, we draw a circle *abca* of radius R about the axis of the solenoid, as in the plan view shown in Fig. 9.1b. The area of the circle *abca* in Fig. 9.1b is πR^2 and the flux of the magnetic field B_z at the time t is given by $F_B = \pi R^2 B_0 \cos(2\pi ft)$. The time rate of change of this flux dF_B/dt is then obtained by differentiation of F_B, and is given by

$$\frac{dF_B}{dt} = -2\pi^2 fR^2 B_0 \cdot \sin(2\pi ft).$$ [9.4]

The strength of the tangential electric field E_T must be the same at all points around the circle *abca* because of the cylindrical symmetry that we have defined for the problem. The field E_T is also always by definition directed tangentially to the circle so that the line integral of E_T around the circle is simply $2\pi RE_T$. Then we may deduce, using Faraday's law, that

$$\text{emf} = 2\pi RE_T(R) = -\frac{dF_B}{dt} = +2\pi^2 fR^2 B_0 \sin(2\pi ft)$$

$$\text{or} \quad E_T(R) = \pi fRB_0 \sin(2\pi ft).$$ [9.5]

Thus we have determined that the application of a magnetic field B_z, which varies sinusoidally with time but is assumed to be uniform in amplitude across the Petri dish at any one time, induces electric fields which form plane closed circles perpendicular to the axis of the solenoid and are sinusoidal with the same frequency as the inducing magnetic field. The amplitude of the sinusoidal tangential electric field is proportional to the amplitude of the magnetic field and to its frequency, and it is also proportional to the radius of the circular path chosen.

If the resistivity of the aqueous solution is ρ_S, then the resistance of the circular ring of solution at a radius R from the axis, which has a length $2\pi R$ and small cross-section α, is given by $2\pi R\rho_S/\alpha$ using equation [3.2]. Applying equation [9.2], the current $I(R)$ circulating in the ring because of the induced emf will be given by the emf divided by the resistance, so that

$$I(R) = \frac{(\text{emf})\alpha}{2\pi R\rho_S}.$$

The current density, or current per unit cross-sectional area, $J(R) = I/\alpha$, circulating in the liquid at a distance R from the axis of the solenoid will be given by

$$J(R) = \frac{\text{emf}}{2\pi R\rho_S} = \frac{2\pi R E_T(R)}{2\pi R\rho_S} = \frac{E_T(R)}{\rho_S}. \qquad [9.6]$$

In the example above, the path *abca* was chosen to have the cylindrical symmetry of the whole experimental set-up such that the induced electric field was always pointing along the path and has a constant amplitude. If we take another path such as the circular path of radius a marked *defd* in Fig. 9.1b then the induced electric field, which forms circles about the solenoid axis, will be parallel to the path in some places, oppositely aligned in others and will in general make an angle to the path. In such a case Faraday's law will still predict the total emf around the complete circular path but will give no detailed information as to the direction or amplitude of the induced electric field at particular positions around the path. Thus we may deduce that the emf around the path *cdec* is given by

$$\text{emf} = +2\pi^2 f a^2 B_0 \sin(2\pi f t). \qquad [9.7]$$

If we do wish to know the amplitude or direction of the induced electric field at a particular position on the path *cdec* then we must use other information such as, in this case, the axial symmetry of the experimental set-up.

9.3   The screening of induced electric fields in biological tissue

With static electric fields we were able to deduce that the equilibrium electric field within a conductor is zero. As described in Section 2.2, this is essentially due to the fact that free charges within the conductor move under the influence of the applied static electric field and accumulate at non-conducting barriers to produce an internal field which opposes the applied field. Thus placing a cell in an external static electric field is not effective in producing electric fields within the cell once equilibrium has been attained. In the case of magnetically induced electric fields which form closed loops, the screening that excludes static electric fields is not possible. These fields may induce currents to flow in the conductor, but there is no accumulation of charge and so no opposing internal electric field is set up. Thus when living cells are exposed to sinusoidal or otherwise time-varying magnetic fields it is likely that electric fields and thus currents will be induced within them. It is for this reason that in many of the experiments designed to test if electric fields within a living cell have a biological effect, the **internal alternating electric fields** are induced by applying an **external alternating magnetic field**.

Such induced emfs within a living cell are, however, limited by the size of the cell. Within a spherical cell of radius a the maximum area perpendicular to the time-varying field that the field can penetrate is πa^2, and the emf is then given by equation [9.7]. For a cell with $a = 10\,\mu m$ and an applied sinusoidal field of frequency 50 Hz and amplitude 10^{-3} T, which is about 20 times the amplitude of the Earth's magnetic field, the amplitude of the maximum possible induced emf is only about 98.7×10^{-12} V.

Knowing the value of the emf around some path within a cell may not be of much value in predicting the induced current because of the inhomogeneous nature of most cells. Internal non-conducting barriers may limit the conducting path, and changes in the resistivity of the cytoplasm along the path may make current prediction problematic.

9.4   The displacement current

The circuit shown in Fig. 9.2 represents a capacitor in the process of being charged by connection through two wires to a battery. Let the current-flowing at some time t be I and let us construct a closed path that circles the current-carrying wire. If we then construct a surface, like that marked 1 in the figure, that is bounded by the path and is cut by the current-carrying conductor, then we would predict using Ampère's law (equation [8.13]) that the line integral of the field $\tilde{H}$ around the path is equal to I, where I is the total free current cutting the surface marked 1 in the figure. If, however, we construct a second surface, like that marked 2 in the figure, which is also bounded by the path but passes between the plates of the capacitor and is not cut by the current, then we would apparently deduce that the line integral of $\tilde{H}$ around the same path is zero because no current cuts this surface. This inconsistency in the application of Ampère's law was first noticed by Maxwell who realized that the flow of some quantity within the capacitor through surface 2 must be equivalent to the flow of the free current through surface 1.

If the capacitor plates have area A and at time t carry total uncompensated free charges of $+Q$ and $-Q$ then the rate of change of this charge with time, dQ/dt, is equal

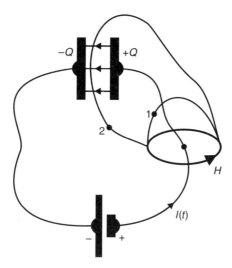

Fig. 9.2   An electrical circuit showing a capacitor in the process of being charged by a battery. At time t the current flowing in the wire is $I(t)$. A path circles around the wire carrying the charging current. Two surfaces marked 1 and 2 are bounded by the same path. Applying Ampère's law to the path and the two surfaces seems to give rise to an anomaly because the charging current passes through surface 1 but not through surface 2. Maxwell deduced from this circuit that a different type of current called the displacement current passes between the plates of the capacitor.

to I, the rate at which charge is delivered to the capacitor. Now the amplitude of the vector $\tilde{D}$ between the plates of the parallel plate capacitor is given by $D = Q/A$ as in equation [5.9], so that within the capacitor

$$\frac{dD}{dt} = \frac{d}{dt}\left(\frac{Q}{A}\right) = \left(\frac{1}{A}\right)\frac{dQ}{dt} = \left(\frac{1}{A}\right)I.$$

Thus the total flux of dD/dt across surface 2 between the capacitor plates of area A, which is given by $A \cdot dD/dt$, is equal in magnitude to I, the total flux of free current through surface 1. Maxwell called dD/dt the **displacement current density** and postulated that Ampère's law should include the flux of the displacement current density $d\tilde{D}/dt$ in addition to the free current density $\tilde{J}_{\text{free}}$ so that it takes the form

$$\oint H_R dR = \int\!\!\int_S \left[\tilde{J}_{\text{free}} + \frac{d\tilde{D}}{dt}\right]_N dS. \tag{9.8}$$

This is the universally applicable form of Ampère's law that has been tested by experiment and found to hold.

Within a good conductor the displacement current density dD/dt is negligible in amplitude in comparison with the free current density J_{free}, except at very high frequencies. However, in regions where there are no conductors, as between the capacitor plates in Fig. 9.2, the displacement current density is of considerable importance even at moderate frequencies. In such regions J_{free} must be zero and equation [9.8] may be written as

$$\oint H_R dR = \frac{d}{dt}\left[\iint_S D_N dS\right]. \tag{9.9}$$

Compare this equation with equation [9.1]; together they show that a time-varying magnetic field gives rise to an electric field and a time-varying electric field gives rise to a magnetic field. At very high frequencies, such as optical frequencies, the displacement current density may dominate the free current density even within a good conductor.

9.5   Maxwell's equations

Here we list the four Maxwell equations that are thought to apply in every situation and that define the properties of electric and magnetic fields:

$$\iint_S D_N dS = \iiint_V \rho_{\text{free}} dV,$$

$$\oint E_R dR = -\frac{d}{dt}\left[\iint_S B_N dS\right],$$

$$\iint_S B_N dS = 0, \tag{9.10}$$

$$\oint H_R dR = \iint_S \left[\tilde{J}_{\text{free}} + \frac{d\tilde{D}}{dt}\right]_N dS.$$

Note that in the second equation the flux of $\tilde{B}$ is taken over an open surface bounded by the path over which the emf is calculated. In the third equation the flux of $\tilde{B}$ is taken over a complete closed surface.

The relationship between the fields $\tilde{D}$ and $\tilde{E}$ is given by

$$\tilde{D} = \varepsilon_0\tilde{E} + \tilde{P} \quad \text{or} \quad \tilde{D} = \varepsilon_R\varepsilon_0\tilde{E}, \tag{9.10}$$

where the first relation is universally true and the second holds for homogeneous dielectrics. Similarly, the relationship between $\tilde{H}$ and $\tilde{B}$ is given by

$$\tilde{B} = \mu_0(\tilde{H} + \tilde{M}) \quad \text{or} \quad \tilde{B} = \mu_R\mu_0\tilde{H}, \tag{9.11}$$

where again the first relation is universally true and the second holds for homogeneously permeable material.

In a region in which there is no free charge ($\rho_{\text{free}} = 0$) and which is non-conducting so that no free currents flow ($J_{\text{free}} = 0$), the equations in electric and magnetic fields show considerable similarities. It is easily shown that these equations combine to predict a self-propagating electromagnetic wave with a velocity v such that

$$v = \frac{1}{\sqrt{\varepsilon_R \varepsilon_0 \mu_R \mu_0}}.$$
[9.12]

or $v = 1/\sqrt{\varepsilon_0 \mu_0}$ in free space. These expressions predict the velocity of light or radio waves which are just examples of the general electromagnetic wave in different frequency ranges. The plane electromagnetic wave is such that the directions of oscillation of $\tilde{E}$ and $\tilde{B}$ are perpendicular to each other and to the direction of propagation. A rotation from the direction of $\tilde{E}$ to the direction of $\tilde{B}$ will advance a right-handed thread in the direction of propagation. Two independent modes are possible for a plane electromagnetic wave travelling in a given direction which have their electric and thus magnetic fields at right angles to each other.

It is beyond the scope of this book to develop the theory of electromagnetic waves but, having shown that an oscillating magnetic field will generate an oscillating electric field and that an oscillating electric field will generate an oscillating magnetic field, it can come as no surprise that together they form a self-propagating wave, even in a vacuum.

Topics covered in Chapter 9

1.   For static fields, which do not vary with time, it is possible to treat electricity and magnetism as two separate subjects. With time-varying fields this is no longer possible as a time-varying magnetic field will give rise to a time-varying electric field and a time-varying electric field will give rise to a time-varying magnetic field.

2.   The time-varying electric field generated by time-varying magnetic fields forms closed loops, unlike the static electric fields which are created by positive charges and terminate on negative charges. The law which characterizes such electric fields is Faraday's law, which states that the line integral of the electric field around any closed path is equal to the negative time rate of change of the flux of the magnetic field that passes through any surface bounded by the chosen path.

3.   Because such electric fields form closed loops, the charges within a conductor which are subject to this type of field will move in closed loops, which does not lead to charge accumulation. Thus the mechanism whereby the charge accumulation within

a conductor subject to an external static electric field leads to zero electric field within the conductor is not effective for electric fields generated by time-varying magnetic fields. Thus an effective way of generating a time-varying electric field within a conductor is to expose it to an external time-varying magnetic field. Within a bounded conductor, such as an animal cell, the maximum area through which the time-varying magnetic field flux may act is limited by the size of the conductor. This limits the amplitude of the maximum current that may be induced within a cell of a given size by a time-varying magnetic field.

4.   Maxwell showed that a displacement current density, given by the time rate of change of the electric displacement $\tilde{D}$, must be added to the free current density in the equation expressing Ampère's law. In regions where there are no conductors the displacement current density is significant at all frequencies. Within a good conductor the displacement current density becomes significant in comparison with the free current density only at very high (optical) frequencies.

5.   The laws that govern the behaviour of electric and magnetic fields can be expressed by four equations in the four parameters E, D, B and H. These, together with the relations that connect D to E and H to B, are called Maxwell's equations.

6.   Essentially because time-varying magnetic fields give rise to time-varying electric fields and vice versa, any time-varying current or charge distribution will give rise to a self-propagating electromagnetic wave. Light, microwaves and radio waves are all examples of this phenomenon. The velocity of these waves in any material may be predicted accurately using only the results of experiments performed on static electric and magnetic fields.

Important equations of Chapter 9

1.   Faraday's law states that the line integral of the electric field $\tilde{E}$ around any closed path is equal to the negative rate of change of the total flux of the magnetic field $\tilde{B}$ through any surface bounded by that path:

$$\oint E_R dR = -\frac{d}{dt}\left[\int\int_S B_N dS\right].$$

The line integral of $\tilde{E}$ around the path is known as the emf generated around that path and is measured in volts.

2.   Electric fields, induced by time-varying magnetic fields, form closed loops. An externally applied time-varying magnetic field can induce circulating electric fields and hence currents within a conductor. The screening effect that ensures that the electric field inside a conductor exposed to a static electric field is zero does not operate

for closed-loop electric fields such as those generated by time-varying magnetic fields.

3.   Ampère's law is completed by the inclusion of the displacement current density $d\tilde{D}/dt$ which can add to the free current density $\tilde{J}_{free}$ in conductors or replace it in insulators where $J_{free} = 0$:

$$\oint H_R dR = \int\int_S \left[\tilde{J}_{free} + \frac{d\tilde{D}}{dt} \right]_N dS.$$

4.   Maxwell's equations defining the properties of electric and magnetic fields may be written as

$$\int\int_S D_N dS = \int\int\int_V \rho_{free} dV,$$

$$\oint E_R dR = -\frac{d}{dt}\left[\int\int_S B_N dS\right],$$

$$\int\int_S B_N dS = 0,$$

$$\oint H_R dR = \int\int_S \left[\tilde{J}_{free} + \frac{d\tilde{D}}{dt}\right]_N dS,$$

$$\tilde{D} = \varepsilon_0 \tilde{E} + \tilde{P} \quad \text{or} \quad \tilde{D} = \varepsilon_R \varepsilon_0 \tilde{E},$$

$$\tilde{B} = \mu_0(\tilde{H} + \tilde{M}) \quad \text{or} \quad \tilde{B} = \mu_R \mu_0 \tilde{H}.$$

Problems

9.1   Calculate the flux of $\tilde{B}$ through a plane coil of area A when the normal to the plane of the coil makes an angle θ to the magnetic field. A magnetic field measuring device consists of a plane circular coil of radius a composed of N turns of wire. The coil has a total resistance R and is rotated about a diameter of the coil at a frequency of f revolutions per second. The rotating coil is inserted between the poles of an electro-magnet such that the axis of rotation is perpendicular to the magnetic field $\tilde{B}$. The amplitude of the sinusoidal alternating current I generated in the coil is measured by an meter of negligible resistance connected across the ends of the coil. Obtain an expression connecting B and the amplitude of I. Assume that the impedance of the coil is purely resistive.

9.2   A model for the magnetic compass for a bird consists of a voltage detector connected to a plane conducting loop of tissue enclosing an area of $10^{-3}\,\mathrm{m}^2$ and located in the wing. Assume that the wing beats on average twice a second and that at the start of each beat the conducting loop has the normal to its plane parallel to the Earth's magnetic field of $50\,\mu\mathrm{T}$. During each beat the loop passes through a position in which the field is in the plane of the loop before returning to its position at the start of the beat. Make a rough estimate of the amplitude of the induced alternating voltage that would register at the detector. Is this a plausible model for a magnetic animal compass?

9.3   A copper ring is placed over the upper end of a vertical soft iron rod on which is wound a solenoid. When a large sinusoidal current is suddenly passed through the solenoid, the ring is observed to spring upward off the end of the rod. Explain this phenomenon.

9.4   Two fixed and parallel conducting rails, separated by a distance L, lie in the horizontal plane. The voltage between the rails is measured by a voltmeter. A conducting rod is laid upon the rails and slides along them with a velocity v such that it remains perpendicular to the rails. If a uniform vertical magnetic field of magnitude B_V exists in the region, find an expression for the voltage registered on the meter using Faraday's law of induction.

9.5   A coil has N turns of closely wound wire carrying a current I. If the flux of the magnetic field passing through the coil due to a single turn is given by ϕ_1, show that the total flux 'threading the coil' is $N^2\phi_1$ and hence that the emf across the coil predicted by Faraday's law is $-N^2(d\phi_1/dt)$. The total flux of the magnetic field threading a coil carrying a current I may be defined as LI. The parameter L is called the **self-inductance** of the coil and depends upon the geometry of the coil and the permeability of material near the coil. Show that the emf across such a coil when it carries a current I is given by $-L(dI/dt)$.

9.6   The conservation of charge requires that the flux of free current J_{free} out of a given closed surface is equal to the rate of diminution of the free charge ρ_{free} stored within that surface, or

$$\iint_S \left[\tilde{J}_{\text{free}}\right]_N dS = -\frac{d}{dt}\left[\iiint_V \rho_{\text{free}}\,dxdydz\right].$$

Combine this expression with Gauss's law in the vector $\tilde{D}$ to prove that the total flux of the quantity

$$\tilde{J}_{\text{free}} + \frac{d\tilde{D}}{dt},$$

which is the sum of the free current and the displacement current, out of any closed surface is zero. This may be seen graphically by imagining a suitable closed surface drawn in Fig. 9.2.

Note that the general form of Ampère's law given in equation [9.8] involves the flux of $\tilde{J}_{\text{free}} + d\tilde{D}/dt$ out of a surface that is not closed but is bounded by the circuit around which the line integral of $\tilde{H}$ is calculated.

10. The motion of a charged particle in electric and magnetic fields and relativity

In this chapter the forces that act upon static and moving charged particles in electric and magnetic fields are collected together as the Lorentz force equation. The motion of a free charged particle and a charged particle subject to a strong centrally directed force are explored as examples of these forces. The Larmor theorem is introduced, which describes in general the rotation that is superimposed on motion that is caused by a strong centrally directed force, when an additional magnetic field is applied. The relativistic approach to the understanding of magnetic fields is briefly discussed and contrasted with the approach taken in this book.

10.1 The Lorentz force

The strength of the magnetic field $\tilde{B}$ was defined in equations [6.1] and [6.2] by the force $d\tilde{F}$ it exerts on a short length dL of conductor carrying a current I such that

$$dF = IBdL\sin(\theta) = |\tilde{I} \times \tilde{B}|dL. \tag{10.1}$$

From this definition we may now calculate the force exerted by the magnetic field on each moving charge. Let the conductor have a cross-sectional area A and let the number of charges q per unit volume within the conductor be N. If the average drift velocity of the charges along the conductor is v, then the number of charges that cross a given cross-section of the conductor per second is NAv. The current I is defined as the charge that crosses a cross-section each second so that $I = qNAv$. The number of charges in the length dL is $NAdL$, which means that the force F_q exerted by the magnetic field on each charge q moving with velocity v is

$$F_q = \frac{dF}{NAdL} = qvB\sin(\theta) = q|\tilde{v} \times \tilde{B}|. \tag{10.2}$$

The force which acts on each charge q is in a direction perpendicular to the velocity $\tilde{v}$ and perpendicular to the magnetic field $\tilde{B}$, as described by equation [6.2] or Appendix 1.

An interesting property of the force exerted on a moving charge by a magnetic field is that it always acts at right angles to the velocity of the charge. This means that the action of a magnetic field can never change the speed of the charge but may change its direction of motion. Note also that a magnetic field exerts no force on a charge that is not in motion. Combining this result with the force $q\tilde{E}$ that is exerted on the charge q by an electric field $\tilde{E}$, we deduce that the total force on the charge in both electric and magnetic fields is

$$\tilde{F} = q[\tilde{E} + \tilde{v} \times \tilde{B}]. \tag{10.3}$$

This is called the **Lorentz force** and it applies universally. We see once again that the electric and magnetic force vectors are $\tilde{E}$ and $\tilde{B}$, and not $\tilde{D}$ and $\tilde{H}$, and that this is true no matter what dielectric or magnetically polarizable material is present. Notice that the velocity v in equation [10.3] is the velocity of the charge **relative to the observer**, which gives a hint that what we call an electric field and what we call a magnetic field may depend upon the motion of the observer. We will return to this topic, which is called relativity, at the end of this chapter.

10.2   The motion of a free charged particle in a static magnetic field

Let us consider a small weight of mass m which is whirling around in a circle because it is tethered to a fixed point by a string of length R. The inertia associated with the weight is associated with a tendency to move in a straight line and it is only the inwardly directed tension in the string that maintains the circular motion. If the speed of the weight around the circle is v, the tension in the string must be mv^2/R to maintain the circular motion. If the string is cut the weight will continue on a straight path that is a tangent to the circle at the position of the weight when the string was cut. The motion of a charged particle in a static magnetic field is similar to our mechanical model with the force exerted by the magnetic field on the charged particle replacing the tension in the string.

In Fig. 10.1 is shown a free particle with mass m and charge q moving with a velocity v anti-clockwise in a circle of radius R about the z-axis and in the (x, y) plane. If a magnetic field $\tilde{B}$ acts along the negative z-axis, into the plane of the figure, an inwardly directed radial force of fixed amplitude qvB will act upon the particle according to equation [10.2]. If the magnetic force equals the centripetal force mv^2/R required for circular motion, then the motion in a circle of radius R will be stable. This will be so if $v = qBR/m$ when the period of the particle rotation becomes τ_c and the frequency f_c such that

$$\tau_c = \frac{2\pi R}{v} = 2\pi\left(\frac{m}{qB}\right) \quad \text{and} \quad f_c = \frac{1}{\tau_c} = \left(\frac{1}{2\pi}\right)\left(\frac{qB}{m}\right). \tag{10.4}$$

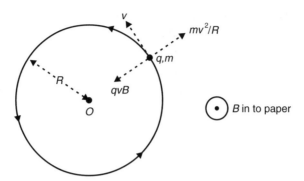

Fig. 10.1   The plane circular path of a particle of mass m and charge q in a magnetic field $\tilde{B}$ directed downwards perpendicular to the plane of the circular path.

The radius does not occur in these formulae, so that all particles have the same period for rotation whatever their radius or speed. Particles with higher speeds circulate in larger circles and those with lower speeds circulate in smaller circles, but **all particles with the same charge-to-mass ratio have the same period of rotation in a given magnetic field**. The common period is called the **cyclotron period**, and the synchronous motion is utilized in particle accelerators such as the cyclotron. The cyclotron frequency formula (equation [10.4]) has also been used for the accurate determination of the charge-to-mass ratio of particles such as the proton and the deuteron circulating in magnetic fields of accurately known amplitude.

If the particle is initially moving in a plane that is not perpendicular to the magnetic field, the velocity of the particle v may be resolved into a component which is in the plane perpendicular to the field v_P, and a component parallel to the field direction v_B. The component v_P leads to cyclotron motion in a circle of radius $R = (m/qB)v_P$ in the plane perpendicular to the field as discussed above. There is no magnetic force caused by the component v_B because this velocity is in the same direction as the field and $\theta = 0$ in equation [10.3]. The resultant motion is cyclotron motion in circles of radius R in the plane perpendicular to the magnetic field, and motion along the field direction at a constant velocity v_B, which results in helical motion with the magnetic field direction as axis.

Note that these relationships apply to a **free particle** which only experiences the effects of an inertial force and a force due to the magnetic field. Some attempts have been made in the biological literature to apply these principles to an ion in aqueous solution. This is clearly totally inappropriate and physically wrong in that the centripetal force has its origin in the inertia of the moving free particle. An ion in solution moves in a viscous regime where its inertia plays a negligible role in shaping its motion, as discussed in Section 3.2. It is in collision with a neighbouring water molecule about ten thousand million times each second, and it is these collisions that determine its motion as a random walk of minute steps.

10.3 The Larmor theorem

In contrast to the **free particle** considered above, we consider here a particle of mass m and charge q under the influence of a **strong centrally directed force**. In Fig. 10.2 we show the motion of a particle of mass m using planar polar coordinates R and θ. As an example of the central force, we will assume that the particle is under the influence of a restoring force which is always directed towards the origin and which has a strength proportional to the distance of the particle from the origin. A situation like this would arise if the particle was connected to the origin by an ideal elastic thread. Under these circumstances the particle will execute **simple harmonic motion** about the origin along the fixed direction of R in the figure. The mass passes forwards and backwards along the direction of R at an angle θ in the figure, passing periodically through the origin. Let us now imagine that a magnetic field is progressively switched on in a direction perpendicular to this plane and pointing down into it. The amplitude of the magnetic field is increased until it reaches a value B and is then held constant. In the Cartesian coordinates shown by the broken lines in Fig. 10.2 the magnetic field $\tilde{B}$ acts along the negative z-axis and into the paper.

We consider the particle leaving the origin and moving out along the direction defined by θ in Fig. 10.2 such that it is at a distance R from the origin at time t. At this

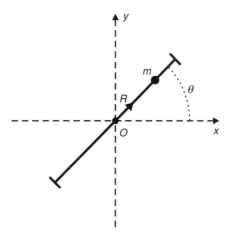

Fig. 10.2   A particle of mass m and charge q executes simple harmonic motion about the origin in the (x, y)-plane in a direction which makes an angle θ to the x-axis. According to the Larmor theorem, if a magnetic field $\tilde{B}$ is applied along the direction of the negative z-axis, the path of the simple harmonic motion will rotate about the z-axis such that $d\theta/dt = (q/(2m))B$.

time it will have an outward radial velocity $v_R = dR/dt$, and because of the magnetic field $\tilde{B}$ it will experience a force $F_B = qv_RB = q(dR/dt)B$ acting in the (x, y) plane and in a direction that tends to increase θ as predicted by equation [10.2]. A force in this direction accelerates the particle within the (x, y) plane and increases the angular velocity $Rd\theta/dt$ of the particle. Continuing its simple harmonic motion along the rotating direction of R, the particle reaches its maximum amplitude and then proceeds inwards towards the origin. In this phase of its motion the radial velocity $v_R = dR/dt$ is negative, and the force due to the magnetic field is in the opposite direction – decelerating the particle in the (x, y) plane and decreasing its angular velocity $Rd\theta/dt$. From this simple description it may be thought that the rate at which the angle θ increases varies with time, perhaps harmonically with the motion of the particle. In fact, as we shall see below, the rate of increase of θ is constant. What determines the motion of the particle is not the magnitude of the force and the angular velocity but the couple about the z-axis created by the force and the angular momentum of the particle.

At time t, when the particle is at a distance R from the origin, it is acted upon by a couple $C(R)$ given by

$$C(R) = RF_B = Rqv_RB = Rq\left(\frac{dR}{dt}\right)B.$$

The angular velocity of the particle at time t is $v_\theta = R(d\theta/dt)$ so that its angular momentum is $Rmv_\theta = RmRd\theta/dt$. The laws of mechanics tell us that the time rate of change of the angular momentum equals the couple that is acting, which as an equation becomes

$$\frac{d}{dt}\left[mR^2\frac{d\theta}{dt}\right] = C(R) = qBR\frac{dR}{dt}. \tag{10.5}$$

Using the mathematical fact that

$$R\frac{dR}{dt} = \frac{1}{2}\cdot\frac{dR^2}{dt},$$

we may rewrite equation [10.5] in the form

$$\frac{d}{dt}\left[R^2\frac{d\theta}{dt}\right] = \frac{qB}{2m}\frac{d}{dt}[R^2].$$

Taking both sides of this equation to the same side, we may rewrite it as

$$\frac{d}{dt}\left\{R^2\left[\frac{d\theta}{dt} - \frac{qB}{2m}\right]\right\} = 0, \tag{10.6}$$

which requires that the quantity

$$R^2 \left[\frac{d\theta}{dt} - \frac{qB}{2m} \right]$$

is a constant which does not change with time. As R is zero periodically, because of the simple harmonic motion, this expression is periodically zero and must therefore remain zero for all time, even when R is finite. This condition requires that the square bracket must always be zero. From this we can deduce that

$$\frac{d\theta}{dt} = \frac{qB}{2m}.$$ [10.7]

Thus we have shown that before the magnetic field was applied the particle performed simple harmonic motion about the origin of the coordinates along a fixed direction R at an angle θ (Fig. 10.2). The effect of the magnetic field is to rotate that direction R, along which the particle oscillated before the field was applied, at a constant angular frequency of $qB/2m$ about the direction of the magnetic field in an anti-clockwise direction.

Because the charged particle now has a small tangential velocity v_θ, perpendicular to the direction of R, caused by the rotation of the direction of R, there will be a small extra force $qv_\theta B$, caused by the presence of the magnetic field, which acts along the direction of R. However, if we assume that this force is negligible in comparison with the original strong centrally directed force which also acts along the direction of R and which is responsible for the harmonic oscillation, then the motion of the particle with the field applied is identical to its original motion except that the whole picture revolves about the direction of the applied magnetic field. The frequency with which this precession occurs is called the **Larmor frequency**, which we write as f_L, and which has a value

$$f_L = \frac{1}{2\pi} \frac{d\theta}{dt} = \frac{1}{2\pi} \frac{qB}{2m}.$$ [10.8]

The Larmor frequency is thus exactly half of the cyclotron frequency of a free particle of the same mass and charge in the same magnetic field.

The result we have obtained in equations [10.7] and [10.8] is one example of the remarkably general Larmor theorem first discovered by Sir Joseph Larmor late in the 19th century. It asserts **that the effect of a magnetic field $\tilde{B}$ on the motion of any particle of mass m and charge q, moving under the influence of a strong centrally directed force, is to superimpose upon the motion, in the absence of the field, a rotation with an angular frequency of $\omega_L = -2\pi f_L = -(qB/2m)$ about the**

direction of the field. The negative sign indicates that the rotation is in the opposite direction to that required to advance a right-hand thread along the direction of the magnetic field. The requirement that the particle is subject to a **strong central force** is in order that secondary forces which arise solely because of the Larmor precession of the charged particle in the magnetic field do not significantly change the original motion, as we discussed above in our example. The effect holds for oscillating particles as in our illustration and also for particles rotating around the origin and attracted towards it by a strong force (see Problem 10.6). It also holds for vibrations and rotations that take place in any plane, and not just the plane perpendicular to the field as in our illustration. Note that in our example the Larmor precession frequency was not dependent upon the frequency of the original harmonic vibration. Thus a charged particle confined within a cavity which executes stochastic vibrations due to random thermal excitation should also experience Larmor precession around the direction of an applied magnetic field. The effect even applies to the motion of the electrons rotating about the nucleus and provides a semi-classical explanation of the Zeeman effect in optical spectroscopy and of diamagnetism, as we discuss below. Probably its most important application is that it forms the basis of our understanding of the semi-classical theory of **magnetic resonance**, which we discuss at length in Chapter 16.

The only forces acting on the particle we considered were the centrally directed force which was responsible for its simple harmonic motion and forces created by the static applied magnetic field. The centrally directed force cannot give rise to angular momentum of the particle and yet, when the field was applied, the particle gained angular momentum about the field direction described by the Larmor precession. We believe that angular momentum is conserved within an isolated system like this, so that we are forced to conclude that there must be angular momentum associated with a static magnetic field. The Larmor precession created by the application of a static magnetic field may be thought of as a direct consequence of the angular momentum inherent in any magnetic field.

10.4   Diamagnetism

If we consider an electron of charge q and mass m circulating about its nucleus in a magnetic field, we can apply the Larmor theorem to deduce the change induced in its motion by the applied field. This predicts an additional angular velocity of $\omega_L = 2\pi f_L = -(qB/2m)$ which takes place in a plane that is perpendicular to the magnetic field $\tilde{B}$, as in equation [10.7]. If at some instant in time the distance of the electron from the nucleus is R, and if the direction of R makes an angle ϕ to the direction of the magnetic field, then the distance of the electron from an axis through the nucleus and parallel to the magnetic field is $R\sin(\phi)$. The tangential Larmor velocity of the electron about the field direction is $v_L = R\sin(\phi)\omega_L$ and the extra angular momentum at this time about the field direction is given by $m[R\sin(\phi)]v_L = m[R\sin(\phi)]^2\omega_L$. Averaged over a complete circuit of the electron around its nucleus, the additional angular momentum becomes

$$dG = -\left(\frac{qB}{2m}\right)m\langle[R\sin(\phi)]^2\rangle_{\text{ave}},\qquad\text{[10.8]}$$

where $\langle[R\sin(\phi)]^2\rangle_{\text{ave}}$ represents the average of this quantity over a complete electron circuit. Throughout this calculation we will assume that the atomic forces are so much stronger than any effect of the applied magnetic field, and that the shape of the orbits remains unchanged in the applied field. Quantum mechanics then tells us that associated with the orbital angular momentum dG of a particle of charge q and mass m there is a magnetic moment with a magnitude dM given by

$$dM = \left(\frac{q}{2m}\right)dG = -\left(\frac{q^2}{4m}\right)\langle[R\sin(\phi)]^2\rangle_{\text{ave}}B.\qquad\text{[10.9]}$$

To find the total induced magnetic moment this expression must be summed over all the electron orbits in the atom. If all angles ϕ are assumed to be equally probable within the atom, it can be shown that

$$\langle[R\sin(\phi)]^2\rangle_{\text{ave}} = \frac{2}{3}\langle R^2\rangle_{\text{ave}}.\qquad\text{[10.10]}$$

Thus the total magnetic moment per atom induced by the magnetic field has amplitude

$$M = -\left(\frac{q^2}{6m}\right)B\langle R^2\rangle_{\text{ave}},\qquad\text{[10.11]}$$

and is directed opposite to the magnetic field. The average is taken over all the electron orbits within the atom.

Note that the induced diamagnetic magnetic moment is proportional to the square of the charge on the particle and would be oppositely directed to the field even if the circulating electrons carried a positive charge. This may be considered as an application of Lenz's law that we introduced in Section 9.1 in that the flux of the applied field that threads the electron orbit induces an emf around the orbit in a direction such as to create an extra magnetic field which opposes the direction of the original inducing magnetic field.

We can make a rough estimate of the diamagnetic induced moment for an atom with atomic number Z, and hence Z electron orbits, if we assume that a typical electron orbit has a radius of 10^{-10} m so that $\langle R^2 \rangle_{\text{ave}}$ is about 10^{-20} m^2. In a field of $B = 50 \times 10^{-3}$ T, which is some 1000 times the amplitude of the Earth's field, the induced magnetic moment per atom becomes $M = -3.3 \times 10^{-30} \cdot Z$ coulomb metres. This to be compared with the permanent paramagnetic moment of about $+1.1 \times 10^{-23}$ C m for a single iron atom. Even allowing for the disturbing effect of the thermal fluctuations on the alignment of the paramagnetic magnetic moments, the positive paramagnetic magnetic moment per unit volume induced by a given magnetic field at room temperature in a paramagnetic solid, is often at least 100 times the size of the negative diamagnetic moment induced in the same material by the same field.

10.5   Special relativity and magnetism

The study of **special relativity** is based upon our belief that the predictions of any satisfactory physical theory should be the same when observed from within two frames of reference moving relative to each other with a constant velocity. Such frames are called **inertial frames of reference**. It is possible to develop a complete electromagnetic theory by studying only electrostatics, as in Chapters 1 to 5, and then applying relativity theory to situations that involve the motion of charges. The predictions of the forces exerted on static and moving charges are then identical to those we have developed in our approach which assumes the presence of both electric and magnetic fields. The relativistic approach is clearly important in a philosophical sense in that it makes the same predictions with a minimum of assumptions, but it is more difficult to apply practically.

It is well known that mass, rate of time passage and distance may appear different to two experimenters observing the same physical situation when the observers are each fixed within a reference frame and when the two reference frames are in relative motion. Fortunately there is abundant experimental evidence that **charge is invariant in all inertial frames**. That is to say that the charge carried by a particle does not depend upon its velocity. It has also been demonstrated that **Gauss's law holds even if the charges within the Gauss law surface are moving**. Thus to two observers moving at a constant relative velocity, the flux of the electric field out of a given surface is always equal to the algebraic sum of the charge within that surface divided by ε_0, even though the size and shape of the surface may appear very different to the two observers. The fact that the total flux of the electric field remains invariant even though the area of the Gauss law surface changes, indicates that amplitudes of electric fields must appear different to the two observers in relative motion. In fact, if an electric field is created by charges static within a given frame of reference, which we will call the **charge frame**, then the same field will appear different to an observer fixed in a second **observer frame** which moves with a fixed velocity relative to the charge frame. However, our fundamental measure of the strength and direction of an electric field $\tilde{E}$ such that the force on a charge q is given by $q\tilde{E}$ still holds provided that the force and the field are measured by an observer from within a single frame of reference. The electric field so measured becomes that relevant to the frame from within which the force was measured. Furthermore, this measure of the electric field holds true no matter what the velocity of the charged particle relative to the observer.

As a further example of special relativity applied to electric fields, we describe the electric field created by a single particle of charge $+q$. As we know, if the particle is static relative to an observer, the electric field radiates uniformly in every direction away from the charge. If the charged particle is moving relative to the observer at a fixed velocity, the electric field remains always directed radially away from the charged particle but the radial uniformity of the field is lost. The faster the velocity of the charged particle relative to the observer, the more the electric field perceived by the observer along the direction of the particle motion diminishes in amplitude, and the more the strength of the electric field at right angles to the direction of motion increases. From the non-relativistic viewpoint of this book there is present in this situation both an electric charge and an electrical current created by the motion of the charge. Thus the observer experiences both a uniformly distributed electric field due to the charge and a magnetic field due to the current. From either viewpoint the total force on a given charge is the same.

From the practical point of view of developing a theory that predicts the outcome of any experiment involving both static and moving charges, the approach we have adopted is easier to visualize and is much less error prone than the relativistic approach. In a sense we have adopted a relativistic approach in stating that static charges give rise to electric fields and that moving charges or currents give rise to magnetic fields. The predicted effects of magnetic fields then encapsulate the properties of charges that arise due to their motion while the predicted effects of electric fields encapsulate the properties of static charges. However, it is as well to be aware of the underlying relativistic nature of magnetism to explain, for example, why the velocity of motion of the charge **relative to the observer** appears in the expression for the Lorentz force in equation [10.3].

As a final example, we show in Fig. 10.3a a beam of particles each with a charge q and moving in the (x, y) plane parallel to the x-axis with a velocity v. A static observer, as in Fig. 10.3(a) observes these charges as forming a current and would thus expect to detect a magnetic field $\tilde{B}$ circling the beam as in equation [7.1]. He would also experience an electric field $\tilde{E}$ in the negative y-direction due to the presence of the charges that constitute the beam. If, however, the same beam of charges is observed by an observer who runs with a constant velocity v in the (x, y) plane parallel to the x-direction as in Fig. 10.3b, then he will observe a linear array of static charges. He

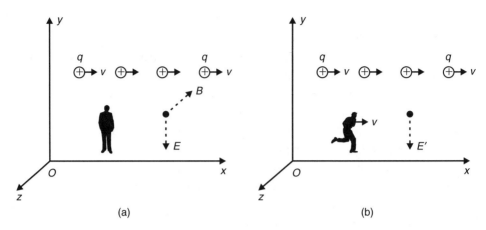

Fig. 10.3 An illustration of the relativistic approach to electricity and magnetism. (a) A static observer in the (x, y)-plane detects a stream of particles with charge q moving in the (x, y)-plane in the x-direction with velocity v. He detects a repulsive electric field along the negative y-axis and a magnetic field circulating about the particle axis because the moving particles constitute a current in the x-direction. (b) If the observer now moves with the same velocity v in the x-direction he observes a static array of charged particles which give rise only to an electric field along the negative y-axis. The observation is altered by the motion of the observer.

will detect no magnetic field but only the electric field in the negative y-direction. In the first case both electric and magnetic fields are observed, while only an electric field is detected in the second case. Here again observation of the same physical situation is changed by the motion of the observer. Without a clear understanding of how we chose to define electric and magnetic fields and the relativistic theory that connects them this example may appear to be paradox.

Topics covered in Chapter 10

1. The forces that act upon a charged particle in electric and magnetic fields are encapsulated in an expression called the Lorentz force. An electric field results in a force in the direction of that field with an amplitude given by the product of the magnitudes of the charge and the field. A magnetic field results in a force in a direction perpendicular to the plane containing the vectors representing the velocity of the particle and the magnetic field. The magnitude of the force is the product of the magnitudes of the charge, the magnetic field, the particle velocity and the sine of the angle between the directions of the particle velocity and the magnetic field. The velocity of the particle is that relative to the observer. A static charge experiences no force in a magnetic field.

2.   In equilibrium in a static magnetic field, a free charged particle moving in a plane perpendicular to the direction of the magnetic field will rotate in circles about the direction of the magnetic field with a frequency known as the cyclotron frequency. All particles with the same charge-to-mass ratio will circulate at the same frequency in a given magnetic field, with faster particles occupying orbits of larger diameter.

3.   The effect of a magnetic field upon the motion of a charged particle subject to a strong centrally directed force is described by the Larmor theorem. This states that the effect of the magnetic field is to superimpose an anti-clockwise rotation of the whole motion about the direction of the magnetic field at a frequency known as the Larmor frequency. The original motion under the strong centrally directed force remains substantially unchanged. All particles with the same charge-to-mass ratio have the same Larmor frequency, which is exactly half the cyclotron frequency that the particle would have in the same applied magnetic field if it were free.

4.   A magnetic field creates in all atoms and molecules a slight distortion of the electron orbits, which results in the generation of a small magnetic dipole moment which points in a direction opposite to the direction of the applied magnetic field. This effect is call diamagnetism.

5.   It is possible to explain all the effects of magnetic fields, which are created by moving charges or currents, by applying the theory of special relativity to the equations that govern static electricity. This approach is important in exploring the structure of the subject but is less suited to the simplest calculation of the effects of moving charges. Our approach has been that static charges create electric fields and that charges in motion, or currents, can give rise to electric and magnetic fields. The relativistic approach does explain why the distinction between electric and magnetic effects, as we have defined them, can depend upon the motion of the observer.

Important equations of Chapter 10

1.   The force acting on a charge q in an electric field $\tilde{E}$ and a magnetic field $\tilde{B}$ is given by

$$\tilde{F} = q[\tilde{E} + \tilde{v} \times \tilde{B}],$$

where v is the velocity of the charge relative to the observer. The force due to the electric field is in the direction of that field. The force due to the magnetic field is perpendicular to the directions of both $\tilde{v}$ and $\tilde{B}$ and has the direction of advance of a right-hand thread rotated from the direction of $\tilde{v}$ to that of $\tilde{B}$.

2.   A free particle of mass m and charge q can form a stable circular orbit in a plane perpendicular to the prevailing magnetic field $\tilde{B}$. The particle circulates at the cyclotron frequency f_C which does not depend upon the radius of the orbit or the speed of the particle. The cyclotron frequency is given by

$$f_C = \frac{1}{2\pi}\left(\frac{qB}{m}\right).$$

3.   The Larmor theorem states that the effect of a magnetic field $\tilde{B}$ on the motion of a particle of mass m and charge q, moving under the influence of a strong centrally directed force, is to superimpose upon the motion in the absence of the field a rotation with a frequency $-f_L$ about the positive direction of the magnetic field, where

$$f_L = \frac{1}{2\pi}\left(\frac{qB}{2m}\right).$$

Problems

10.1   A horizontal conducting rod of length L is moved in a direction perpendicular to its length with velocity v in a region where there exists a vertical magnetic field of strength B. Show that the charge carriers in the rod move under the influence of the Lorentz force until a voltage $V = vBL$ is established across the ends of the rod. Note that this deduction does not rely upon Faraday's law but only the Lorentz force, but that it leads to the same conclusion, as is shown in Problem 9.4.

10.2   An inventor plans to measure the vertical component of the Earth's magnetic field B_V by attaching an insulated straight conducting rod of length L above the wings of an aeroplane and then flying horizontally at a speed v. What is the voltage V_{rod} induced across the ends of the rod? The inventor intends to measure this voltage by means of two wires attached to the ends of the rod and leading to a voltmeter in the cockpit. Deduce, giving your reasons, the voltage that will register on the voltmeter. In Section 14.3 we describe how aquatic animals may detect the amplitude of the vertical component of the Earth's magnetic field using an ingenious technique based upon induction.

10.3   A beam of charged particles moves along the direction of the y-axis with a velocity v. An electric field of magnitude E, applied along the x-direction, deflects the particles. Find the magnitude and direction of a magnetic field, applied together with the electric field, that will restore the original undeflected motion of the particles. Draw a diagram of the Cartesian axes showing the directions of $\tilde{v}$ and the electric and magnetic fields and forces.

10.4   A free proton with charge 1.6×10^{-19} C and mass 1.67×10^{-27} kg is in stable motion in plane circles in a magnetic field with magnitude 1 T directed normally to the circle. Estimate the cyclotron resonance frequency of the proton. What is the velocity of the particle if the radius of the circle is 0.1 m?

10.5   A free particle of mass m and charge q moves anti-clockwise in circles of radius R in the (x, y) plane as in Fig. 10.1. The stable motion is cyclotron resonance in a magnetic field B_0 directed into the plane of the figure. The magnetic field is slowly increased in amplitude until it reaches $B_0 + \Delta B$. As the field increases the particle is observed to accelerate. Show, using Faraday's law, that the particle is acted upon by an electric field which accelerates it during the period in which the magnetic field increases. Deduce a relationship between the velocity and the radius of the motion before and after the magnetic field is applied. How did the period of the rotation change when the magnetic field was increased?

10.6*   A particle of charge q and mass m is attached to the origin of Cartesian coordinates by a light inelastic string of length R_0. The particle rotates rapidly with speed v_0 in a circle of radius R_0 with angular frequency w_0, in the (x, y) plane and in an anti-clockwise direction when looking down the z-axis. A magnetic field of magnitude B is now applied along the direction of the negative z-axis. According to the Larmor theorem, the effect of the field is to increase the angular frequency of rotation by an amount $\Delta w = qB/2m$ but to leave the system otherwise unchanged provided that $w_0 \gg \Delta w$. Show that, provided $w_0 \gg \Delta w$, the extra centripetal force in the string caused by the increase in angular frequency from w_0 to $w_0 + \Delta w$ is balanced by a new inward-directed force due to the presence of the increased magnetic field.

Worked solution

In the absence of the magnetic field, $v_0 = R_0 w_0$ and the outward centripetal force F_c, which is balanced by tension in the string, is given by

$$F_c = \frac{mv_0^2}{R_0} = mR_0 w_0^2.$$

When the magnetic field is applied, the angular velocity increases from w_0 to $w_0 + \Delta w$, where Δw is given by the Larmor theorem as

$$\Delta w = \frac{qB}{2m}.$$

The centripetal force becomes $F_c = mR_0(\omega_0 + \Delta\omega)^2$ so that the increase in the outward-directed centripetal force is given by

$$\Delta F_c = mR_0(2\omega_0\Delta\omega + \Delta\omega^2).$$

The extra inward force caused by the magnetic field ΔF_B is equal to $qB(v_0 + \Delta v) = qBR_0(\omega_0 + \Delta\omega)$. Substituting for B from the Larmor expression for $\Delta\omega$, we find that

$$\Delta F_B = q\left(\frac{2m\Delta\omega}{q}\right)R_0(\omega_0 + \Delta\omega) = mR_0(2\omega_0\Delta\omega + 2\Delta\omega^2).$$

Thus, provided that $\omega_0 \gg \Delta\omega$ so that the second terms in the expressions for ΔF_c and ΔF_B are very small in comparison with the first terms, we have shown that

$$\Delta F_c = \Delta F_B.$$

Note that as a result of this equality the string would remain the same length even if it were elastic, as would also be predicted by the Larmor theorem. Such an elastic string could represent the electrical attraction between the rotating charge and a charge of opposite sign at the origin and our model could then represent semiclassically the circulation of an electron about a nucleus. Can you give a direct physical explanation for the increase in the angular velocity while the magnetic field is being applied, using Faraday's law?

PART II: APPLICATIONS

11. Ions in aqueous solution and the ionization of acids and bases

The behaviour of ions in aqueous solution is of paramount importance in biology and serves as an application of many of the equations that govern electrical behaviour that have been derived in previous chapters. The topics discussed in this chapter include the stability of the equilibrium state, electrical neutrality, pH and pK_a and the influence of electric fields and dielectric environment on these parameters.

11.1 Ions in aqueous solution

The properties of ions in bulk solution are largely determined by equilibrium thermo-dynamics. As discussed in Appendix 2, any deviations from the equilibrium state that do occur are likely to be small. In addition to the fact that the deviations are small, the kinetics of the motions that restore the equilibrium are fast. Let us consider an aqueous solution of potassium chloride with a concentration of 0.1 molar as a much simplified model of a biological solution. The mean separation of the cations is 2.55×10^{-9} m as each cation occupies an average volume of 1.66×10^{-26} m^3. Using equation [3.4], we deduce that at 25 °C the average time taken for a potassium ion to thermally diffuse this distance is only 5.5×10^{-10} s. Thus the solution will correct any small, and thus probable, non-equilibrium distribution of ions very rapidly.

A second equilibrium property that we can rely upon in bulk solution is electrical neutrality. Electrostatic forces are so strong and the number of ions per unit volume so large, that any local deviation from electrical neutrality in a solution will generate large electric fields that drive neighbouring ions to correct any deviation from neutrality. If within some region of our model solution of 0.1 M potassium chloride the number of cations was greater than the number of anions by only 1%, the resulting charge density would be 9.63×10^4 C m^{-3}, which is very large. The electric fields generated by this excess charge density will drive cations from the region and attract anions into the region to remove the deficit of anions. As another hypothetical example, if we were to effectively remove the charge of a single cation by placing an equal charge of opposite sign upon it, the electric field at the average near-neighbour cation sites in our previously neutral solution would be about $2.8 \times 10^6 \, V \, m^{-1}$ attracting them toward the charge vacancy. With electric fields of this size operating, the neutrality of the solution would be rapidly restored. Considerations such as these lead us to believe that bulk solutions in equilibrium are highly homogeneous and that we can rely upon average equilibrium properties being accurately maintained.

As discussed briefly in Section 5.1, the centre of the distribution of the negative charge on the oxygen atom of a water molecule is displaced from the centres of the distributions of positive charge of the hydrogen atoms, giving rise to a permanent electric dipole moment of the water molecule. An important consequence of this charge distribution is that a water molecule can act as a donor and an acceptor of hydrogen bonds simultaneously. Very few small molecules have this property. A water–water hydrogen bond can to a first approximation be considered as the electrostatic attraction between the electric dipole moment of the electrically polarized OH orbital of one molecule and a negatively charged 'lone pair' orbital of the oxygen molecule of the other molecule. More accurately, there is a slight covalent character to the bond because of some overlap of the two orbitals. The hydrogen bond is directional and has its greatest strength when the two orbitals are aligned along a single axis. A single water molecule can thus form four tetrahedrally oriented hydrogen bonds with neighbouring water molecules. Two of the bonds are along the electrically polarized OH orbitals towards the lone pair oxygen orbitals of neighbouring water molecules, and for these two hydrogen the original molecule is known as a hydrogen bond **donor**. The molecule also acts as an **acceptor** of two hydrogen bonds donated by neighbouring water molecules which are directed towards its two oxygen lone pair orbitals. The open, hydrogen-bonded tetrahedral array of water molecules is clearly seen in the crystal structure of ice as shown in Fig. 11.1 and accounts for the fact that water expands on freezing. The drop in energy, or more strictly enthalpy, on forming a single hydrogen bond between two water molecules in water is thought to be about $3 \times 10^{-20} \, J$. This enthalpy drop is considerable bigger than a typical thermal energy kT which is about $4.1 \times 10^{-21} \, J$ at $25 \, °C$. Thus even in liquid water, the molecules are thought to form dynamic structures that maintain as many of the four possible hydrogen bonds with neighbouring molecules as is possible, even although the half-life of a given hydrogen bond in water can be as short as $10^{-11} \, s$. The tendency to order by hydrogen bonding is opposed by the tendency towards disorder and an increase in entropy as discussed in Appendix 2. However, in this case, as is typical when dealing with strong electrostatic interactions, the drop in the Gibbs free energy (see Appendix 3) on hydrogen bonding due to the enthalpy term is some 50 times as large as the rise in the free energy due to the entropy term at $25 \, °C$.

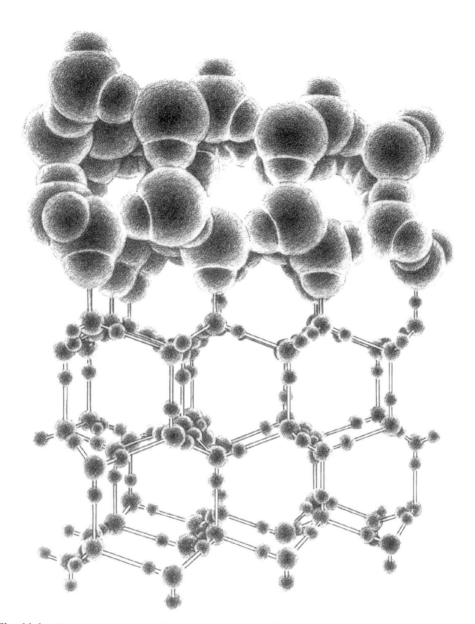

Fig. 11.1   The crystal structure of normal hexagonal ice. Note the open structure, with each molecule acting as a donor of two hydrogen bonds and an acceptor of two hydrogen bonds from adjacent molecules. COLLEGE CHEMISTRY © 1964 by Linus Pauling. Used with the permission of W. H. Freeman and Company.

Water has a concentration of 55.6 molar, so that even with an ionic concentration of 0.1 molar, which is towards the high end of the range found in living systems, there are over 500 water molecules for every cation or anion. As discussed in Section 5.1, water molecules have a large electric dipole moment p of magnitude 6.14×10^{-30} C m which bisects the HOH angle between the two OH sounds and points away from the oxygen atom. From equation [4.5] we see that in an electric field $\tilde{E}$ these dipoles have an energy $-pE\cos(\theta)$, where θ is the angle between the directions of the electric field and the dipole moment. Thus the electrostatic energy of an electric dipole is lowest in a region where the electric field is largest and when the dipole moment is aligned with the electric field. The radial electric field at the surface of small ions is so great that it is capable of orienting and immobilizing the water molecules that touch it, despite the consequent reduction in hydrogen bonding between these water molecules and other water molecules. These structures of immobilized water molecules are sufficiently long-lived to be detected by X-ray diffraction (Enderby and Neilson 1979) and are characteristic of the type of ion. The next layer of water from the ion will have a time-averaged electric dipole moment partially aligned in the radial electric field, but the thermally excited motion of the tumbling water molecules is too fast to show a structure by X-ray or neutron diffraction. A small ion carries this **hydration shell** of ordered and partially ordered water about with it in aqueous solution. The presence of the hydration shell of water carried along by a small ion in aqueous solution is the reason why a small ion can be less mobile than bigger ions in aqueous solution, as reflected in the listing of some ionic diffusion constants in Table 3.1.

The lowering of the electrostatic self-energy $\frac{1}{2}qV_S$ of a hydrated ion with charge q, discussed in Sections 5.5 and 5.8 on a macroscopic scale as being a consequence of the large dielectric constant of water, may be thought of on a microscopic scale as due to the lowering of the voltage at the surface of the ion V_S by the electric fields of the ordered and partially ordered electric dipole moments of the water molecules that surround the ion.

11.2   The dissociation of the water molecule and pH

Pure water dissociates very slightly to form OH^- and H_3O^+ ions. A chemical equation may be written as

$$2H_2O \underset{k_2}{\overset{k_1}{\rightleftharpoons}} OH^- + H_3O^+ \qquad [11.1]$$

where k_1 and k_2 are rate constants. In equilibrium, the law of mass action requires that

$$[H_2O] \cdot [H_2O] \cdot k_1 = [OH^-] \cdot [H_3O^+] \cdot k_2, \qquad [11.2]$$

where the square brackets represent the **molar concentrations** of the constituents. The law of mass action is simply a statement that for reaction [11.1] to proceed to the right requires the simultaneous presence at one location of two water molecules, which is proportional to $[H_2O] \cdot [H_2O]$, multiplied by the rate constant k_1, with a similar product for the reaction to proceed to the left. In equilibrium the rates of reaction in the two directions must be equal. As everywhere in these simplified discussions, all the solutions are supposed to be sufficiently dilute so that activities are replaced by the molar concentrations. In fact the ionic concentrations considered here are very small, so that $[H_2O]$ may be taken as a constant. It is found experimentally that at 25 °C

$$[OH^-] \cdot [H_3O^+] = 1.008 \times 10^{-14}. \qquad [11.3]$$

The concentrations $[OH^-]$ and $[H_3O^+]$ vary over a very large range, so that a logarithmic scale is used. They are defined in terms of a parameter pH such that

$$pH = -\log_{10}([H_3O^+]) \quad \text{or} \quad [H_3O^+] = 10^{-pH}. \qquad [11.4]$$

Thus in an electrically neutral water solution when $[H_3O^+] = [OH^-]$, each concentration must be close to 10^{-7} m by equation [11.3]. Therefore from equation [11.4] the pH is very close to 7. A low pH signifies a high value for $[H_3O^+]$, and a high pH represents a low value of $[H_3O^+]$. The value of $[OH^-]$ is determined from equation [11.3], once the pH is known, because $[OH^-] = 10^{(pH-14)}$ to good accuracy.

   The ions H_3O^+ and OH^- will of course carry with them in aqueous solution a hydration shell of polarized and partially polarized water molecules in the manner we discussed above for metal ions. The hydrated proton, which is represented by H_3O^+ or sometimes simply by H^+, and the hydrated hydroxyl ion have particularly high diffusion constants because the proton can move by tunnelling along a hydrogen bond between adjacent water molecules without requiring translation of the water molecules. Thus a proton may tunnel along the hydrogen bond connecting a H_3O^+ molecule and a neighbouring water molecule and thus convert the H_3O^+ ion to an uncharged water molecule while creating a H_3O^+ ion at the position of the original neighbouring water molecule. The importance of this mechanism may be judged by the fact that the diffusion constant of the proton in pure ice, where the water molecules may not translate, is higher than that of the proton in water where water molecules are free to translate. The high diffusion constant of the proton in water, $D = 9.31 \times 10^{-9}$ m^2 s^{-1} means that it often plays an important role in biology.

11.3   Ionizable residues

Most of the charged groups in proteins consist of ionizable acid or basic residues. For these purposes we define an acid as a proton donor and a base as a proton acceptor, although more general definitions exist. The chemical equation representing the dissociation of an acid AH to its conjugate base A^- may be written as

$$AH + H_2O \underset{k_2}{\overset{k_1}{\rightleftharpoons}} A^- + H_3O^+ \qquad [11.5]$$

where k_1 and k_2 are rate constants. In equilibrium, the law of mass action requires that

$$[AH] \cdot [H_2O] \cdot k_1 = [A^-] \cdot [H_3O^+] \cdot k_2, \qquad [11.6]$$

where the quantities in square brackets again represent molar concentrations. The acid dissociation constant is then defined by K_D in the expression

$$K_D = \frac{k_1}{k_2} = \frac{[A^-] \cdot [H_3O^+]}{[AH] \cdot [H_2O]}. \qquad [11.7]$$

However, as $[H_2O]$ is essentially constant, we can define a new constant K_a which characterizes the dissociation so that

$$K_a = \frac{[A^-] \cdot [H_3O^+]}{[AH]}. \qquad [11.8]$$

Because K_a varies over such a large range of values, the dissociation of the acid is characterized by a number pK_a on a logarithmic scale, such that

$$pK_a = -\log_{10}(K_a) \quad \text{or} \quad K_a = 10^{-pK_a}. \qquad [11.9]$$

Substituting the value of $[H_3O^+]$ from equation [11.4] into equation [11.8] and then using equation [11.9], we obtain the ratio of the number of charged residues to the number uncharged as

$$\frac{[A^-]}{[AH]} = \frac{K_a}{[H_3O^+]} = \frac{10^{-pK_a}}{10^{-pH}} = 10^{(pH-pK_a)}. \qquad [11.10]$$

From equation [11.10] it is seen that the absolute probability of finding the acid site in its negatively charged ionized state $P_a(-)$ becomes

$$P_a(-) = \frac{[A^-]}{[A^-] + [AH]} = \frac{1}{1 + \frac{[AH]}{[A^-]}} = \frac{1}{1 + 10^{(pK_a-pH)}}. \qquad [11.11]$$

Note that this probability only depends upon the difference between pH and pK_a and not upon their individual values. When $pH = pK_a$ the value of $P_a(-)$ is 0.5 and the probabilities of finding the residue charged and uncharged are equal. Positive values of $pK_a - pH$ mean that $P_a(-)$ is less than 0.5 and negative values of $pK_a - pH$ mean that $P_a(-)$ is greater than 0.5. This is what we would expect because a larger pH corresponds to a smaller local concentration of hydrated protons and a lower probability that the acid residue is protonated and thus a higher probability that it is negatively charged.

For every acid there exists a conjugate base as in equation [11.5] and to every base there corresponds a conjugate acid. When dealing with the binding of a proton to a basic residue B, it is conventional to describe the reaction in terms of the acid dissociation constant pK_a of its conjugate acid BH^+ rather than define another proton binding constant for the basic residue itself. The chemical equation for the protonization of a base may be written

$$B + H_3O^+ \underset{k_2}{\overset{k_1}{\rightleftharpoons}} BH^+ + H_2O. \qquad [11.12]$$

Following the same arguments that were used for the acid AH above, the pK_a for the dissociation of the acid BH^+ becomes

$$pK_a = -\log_{10}(K_a) \quad \text{where} \quad K_a = \frac{k_2}{k_1} = \frac{[B] \cdot [H_3O^+]}{[BH^+]}. \qquad [11.13]$$

Again following the arguments that we used for the acid, we deduce that

$$\frac{[BH^+]}{[B]} = 10^{(pK_a - pH)}$$

$$\text{and} \quad P_b(+) = \frac{[BH^+]}{[B] + [BH^+]} = \frac{1}{1 + \frac{[B]}{[BH^+]}} = \frac{1}{1 + 10^{(pH - pK_a)}}. \qquad [11.14]$$

where $P_b(+)$ is the probability of finding the base in its positively charged state.

11.4   The effects of electric fields on the ionization of acid and basic residues

The probability of finding an ionizable residue in its ionized state is changed by the presence of an externally applied voltage because the voltage changes the local concentration of protons $[H_3O^+]_{local}$. If the concentration of protons in the bulk solution, where the externally applied voltage is zero, is given by $[H_3O^+]_{bulk}$ and the concentration of protons in a local region where there exists an applied voltage V is given by $[H_3O^+]_{local}$, then the Boltzmann distribution function as described in Appendix 2 predicts that

$$[H_3O^+]_{local} = [H_3O^+]_{bulk} \exp\left(\frac{-qV}{kT}\right), \qquad [11.15]$$

where q is the charge of the proton and k is the Boltzmann constant. Here qV is the increase in energy of the proton when it moves from the bulk solution where $V = 0$ to the local region where the voltage is V. If V is positive this voltage will tend to drive positively charged protons from the local region and thus reduce $[H_3O^+]_{local}$. When determining the ratio $[A^-]/[AH]$ it is the local concentration of protons near the site that is important. Thus equation [11.10] becomes

$$\frac{[A^-]}{[AH]} = \frac{K_a}{[H_3O^+]_{local}} = \frac{10^{-pK_a}}{10^{-pH}\exp\left(\frac{-qV}{kT}\right)} = 10^{(pH-pK_a)}\exp\left(\frac{+qV}{kT}\right).$$

Because of the change in this ratio the probability $P_a(-)$ that an acid residue is in its charged state becomes

$$P_a(-) = \frac{1}{1 + \frac{[AH]}{[A^-]}} = \frac{1}{1 + 10^{(pK_a - pH)}\exp\left(\frac{-qV}{kT}\right)}. \qquad [11.16]$$

Care must be taken when applying equation [11.16] that the voltage caused by the residue itself, in its negatively charged state, is not included in V because the effect of this voltage on neighbouring protons is already included in the definition of pK_a.

These voltage effects explain the fact that two identical acid residues that are close together on a protein may behave very differently. When both residues are uncharged there exists no externally applied voltage and the value of $P_a(-)$ is given by equation [11.11]. When the first residue is negatively charged, the second residue experiences a negative voltage V and the probability $P_a(-)$ that the second residue becomes charged is determined by equation [11.16] where V is the negative voltage on the second residue due to the first charged residue. The second residue is now much less likely to change to its charged state because the negative voltage due to the first residue has increased markedly the concentration of protons $[H_3O^+]_{local}$ near the second residue. Although this effect is often described in textbooks by stating that the 'effective pK_a' of the second residue has changed it is in fact the 'effective pH' near the second residue that has changed according to equation [11.15]. Note that the negative voltage at the second residue due to the first is an externally applied voltage when seen from the viewpoint of the second residue and thus does rightly appear in equations [11.15] and [11.16].

11.5 The effects of the electrical polarizability of the environment on the ionization of residues

To assess the effect of the material environment on the pK_a of an ionizable residue we must first obtain a relationship connecting the dissociation constant for the ionization and the change in Gibbs free energy that results from the dissociation. In Section A3.4

we show that the change in the Gibbs free energy of a system at constant temperature and pressure may be written in terms of changes in the numbers of the constituents and their electrochemical potentials, so that we have

$$dG_{T,p} = \sum_i \mu_i dn_i. \tag{11.17}$$

In this equation dn_i represents the change in the number n_i of the ith type of constituent which has an electrochemical potential in this system of μ_i. If we write the electrochemical potential for particles of type i per mole rather than per particle as in Appendix 3, we obtain

$$\mu_i = \mu_{0i} + NqV + NkT \log_e \left(\frac{C_i}{1}\right), \tag{11.18}$$

where μ_{0i} is the electrochemical potential in the standard state, N is Avogadro's number, C_i is the molar concentration and the 1 represents a concentration of 1 molar. Applying equation [11.17] to the dissociation of an acid as depicted in equation [11.5] when dn moles of the acid are involved, we obtain

$$dG_{T,p} = dn \left\{ \mu_0(A^-) + \mu_0(H_3O^+) - \mu_0(AH) - \mu_0(H_2O). \right.$$
$$\left. + NkT \log_e \left(\frac{[A^-] \cdot [H_3O^+]}{[AH] \cdot [H_2O]}\right) \right\}, \tag{11.19}$$

where the square brackets represent the molar concentrations as in equation [11.6]. To obtain this result we have assumed that the voltage V is everywhere zero and we have used the fact that

$$\log\left(\frac{AB}{CD}\right) = \log(A) + \log(B) - \log(C) - \log(D).$$

The first four terms in equation [11.19], which form the signed sum of the molar electrochemical potentials in the standard state, we write as $dG^0_{T,p}$. The argument of the logarithm is just the acid dissociation constant K_D as in equation [11.7] so that we may substitute this into equation [11.19] to obtain

$$dG_{T,p} = dn \left[dG^0_{T,p} + NkT \log_e(K_D) \right]. \tag{11.20}$$

For a system in equilibrium, we know from Appendix 3 that $dG_{T,p}$ is zero. Thus from equation [11.20] we obtain

$$dG^0_{T,p} = -NkT \log_e(K_D) \quad \text{or} \quad K_D = \exp\left(\frac{-dG^0_{T,p}}{NkT}\right), \tag{11.21}$$

which is the required relation connecting the dissociation constant and the change in the Gibbs free energy in the standard state. This has the expected form because a large negative value of $dG_{T,p}^0$ represents a spontaneous favourable reaction that is also represented by a large value of the dissociation constant K_D. Strong acids that lead to almost complete dissociation are represented by large values of K_D, as can be seen from equation [11.7]

Let us suppose that we are dealing with the dissociation of an acid residue AH on the surface of a protein which is exposed to the surrounding aqueous solution with a known pH. In this location we measure its pK_a and thus deduce the probability that it is charged using equation [11.11]. We now wish to know what would be the pK_a of an identical residue buried inside the hydrophobic core of the protein rather than being exposed on its surface. One way of solving this problem (Warshel and Russell 1984) is to perform the three-step thermodynamic cycle shown in Figure 11.2.

The first step of the cycle involves removing the hydrated acid residue AH from its position within the protein to the position on the protein surface where the pK_a is

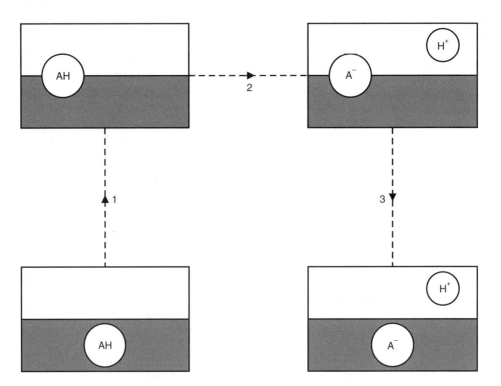

Fig. 11.2 A thermodynamic cycle of three steps. Step 1 shows the transfer of an acid AH from the interior of a protein to its surface which is bathed with aqueous solution. Step 2 shows the ionization of the acid to A$^-$ and a proton H$^+$. Step 3 shows the return of A$^-$ to its original position while the proton remains in the bathing solution.

known. The second step involves the dissociation of the acid at its surface site. The third step involves returning the ionized residue A^- to its original position but leaving the hydrated proton, H_3O^+ or H^+ in the aqueous solution. The change in the Gibbs free energy resulting from a transition from the initial state with AH located in the interior of the protein to a final state with the ionized residue in this position but the dissociated hydrated proton translated to the surrounding solution, is given by the sum of the Gibbs free energy changes in the three steps of the cycle.

The change in the Gibbs free energy in the second step when dissociating one mole of the acid on the surface is given by equation [11.21] with K_D determined by the measured values of pH and pK_a through equations [11.7], [11.8] and [11.9]. There is a Gibbs free energy change in step 1 but this is overwhelmed by the rise in the Gibbs free energy in step 3 due to the large rise in electrostatic self-energy when the charged residue is moved in step 3 from an aqueous environment with high relative dielectric constant to the hydrophobic core of the protein with a low relative dielectric constant, as we discussed in Section 5.8. The rise in the Gibbs free energy change over the three steps compared to that in the second step alone leads, through equation [11.21], to a lower value of K_D for the acid dissociation within the protein in comparison with that for the same residue exposed on the surface. This thermodynamic cycle shows that the same ionizable residue is much less likely to be found in its charged state when its environment has a low electrical polarizability than when it is exposed to a highly polarizable environment. It is for this reason that charged residues are seldom found within the hydrophobic core of a protein or within a lipid bilayer.

Some simple examples of the predicted changes that will occur in the probability of an ionizable residue being charged when the external electric field at the site of the residue or its dielectric environment changes are given in Edmonds (1989).

References

Edmonds, D. T. (1989) A kinetic role for ionizable sites in membrane channel proteins. *European Biophysics Journal* **17**, 113–119.

Enderby, J. E. and Neilson, G. W. (1979) X-ray and neutron scattering by aqueous solutions of electrolytes. In: *Water: A Comprehensive Treatise* (Ed. F. Franks), Vol. 6, Chapter 1. Plenum Press, London.

Warshel, A. and Russell, S. T. (1984) Electrostatic interactions in biological systems and solution. *Quarterly Review of Biophysics* **17**, 283–422.

12.   The Debye layer

Many of the membranes found in biological cells are charged, and this has a profound effect upon the concentration of ions to be found in their vicinity. The layer of solution close to such charged surfaces is known as the Debye layer and many of its properties are discussed in this chapter, including the magnitude of the electric field when the surface charge density is known and the anomalous ionic concentrations that result from the presence of this electric field. The validity of some of the assumptions that have to be made in a simple theory of the Debye layer are discussed.

12.1   The basic electrostatics

Many surfaces within biological cells are charged, and this has a profound effect upon the concentration of ions in their vicinity. The region near the charged surface where these effects operate is called the **Debye layer**. Figure 12.1 shows a large plane surface carrying a surface charge of $+\sigma_S$ per unit area bathed by an aqueous ionic solution on the right. Some negatively charged counterions from the bathing solution will be so strongly attracted to the positively charged surface as to be immobilized on that surface. This narrow layer is called the **Stern layer**, and we will include any permanently attached ions as part of the charged surface. We therefore define the distance x, perpendicular to the surface, as starting on the outer face of the Stern layer. From the planar symmetry of the problem, we know that the electric field $E(x)$, the electric displacement $D(x)$ and the voltage $V(x)$ are constant in planes perpendicular to the x-axis and vary only with x, and that the direction of the electric field $\tilde{E}$ and the electric displacement $\tilde{D}$ are directed perpendicularly to the surface and outward, for positive σ_S.

We construct a Gauss law surface in the form of a circular cylinder with its axis perpendicular to the charged surface and with a length dx and a cross-section of area A as in Fig. 12.1. Let its two flat ends be located at distances x and $x + dx$ from the charged surface. Because the curved sides of the cylinder are parallel to $\tilde{D}$, the total flux of D out of the cylinder is only through its flat ends and is given by $A[D(x + dx) - D(x)] = A[dD(x)/dx]dx$, where $dD(x)/dx$ is the rate of change of $D(x)$ with x. If the **free charge** per unit volume in the bathing solution at x is given by $\rho_F(x)$,

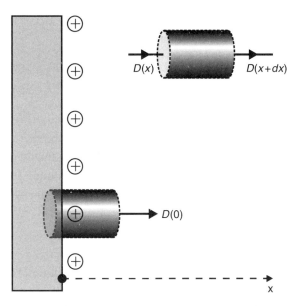

Fig. 12.1 A plane membrane with positive surface charge bathed on the right by an aqueous solution of ions. Two Gauss law circular cylinders are shown with their axes normal to the membrane surface. One has its plane ending at positions given by x and $x + dx$. The second short cylinder is buried in the surface of the membrane.

then Gauss's law as in equation [4.14] requires that

$$A \left[\frac{dD(x)}{dx} \right] dx = \rho_F(x) A dx \quad \text{or} \quad \frac{dD(x)}{dx} = \rho_F(x). \qquad [12.1]$$

If the relative dielectric constant of the fluid bathing the charged surface is ε_R, we may write $D(x) = \varepsilon_R \varepsilon_0 E(x)$. We may also, using equation [1.9], write $E(x) = -dV(x)/dx$. Substituting these two relations into equation [12.1], we obtain

$$\frac{d^2V(x)}{dx^2} = \frac{d}{dx}\left[\frac{dV(x)}{dx}\right] = \frac{d}{dx}[-E(x)] = -\frac{1}{\varepsilon_R \varepsilon_0} \cdot \frac{dD(x)}{dx} = \frac{-\rho_F(x)}{\varepsilon_R \varepsilon_R}. \qquad [12.2]$$

Let us assume that, far away from the charged surface, where $V(x) = 0$, the number of ions of type i per unit volume is given by n_i. Let the valence of this type of ion be z_i so that its charge is $z_i q$, where q is the proton charge. Note that z_i can have either sign, so that for univalent cations and anions it is 1 and -1 respectively and for divalent cations and anions it is $+2$ and -2 respectively. In a region where the voltage is $V(x)$, the energy of an ion of type i is raised by $z_i q V(x)$, using the definition of the voltage given in Section 1.6. We see in Appendix 2 that the ratio of the probability $P(U)$ of finding a particle where its energy is U to the probability $P(0)$ of finding it where its energy is 0 is given by the Boltzmann factor

$$\frac{P(U)}{P(0)} = \exp\left(-\frac{U-0}{kT}\right) = \exp\left(\frac{-U}{kT}\right).$$

Thus the number of ions of type i per unit volume in a location where the voltage is $V(x)$ is given by $n_i \exp(-z_i q V(x)/kT)$ because n_i is the number per unit volume where the voltage is zero. In these expressions k is the Boltzmann constant equal to $1.38 \times 10^{-23}\,\mathrm{J\,K^{-1}}$ and T is the absolute temperature. Summing over all i to include all types of ion, we deduce that the free charge per unit volume at x is given by

$$\rho_F(x) = \sum_i z_i q n_i \exp\left(\frac{-z_i q V(x)}{kT}\right). \tag{12.3}$$

Substituting the value of ρ_F from equation [12.3] into equation [12.2], we obtain the second-order differential equation in $V(x)$,

$$\frac{d^2 V(x)}{dx^2} = -\sum_i \left(\frac{z_i q n_i}{\varepsilon_R \varepsilon_0}\right) \exp\left(\frac{-z_i q V(x)}{kT}\right). \tag{12.4}$$

This is the equation that allows us to determine the variation of the voltage $V(x)$ with the distance x. In fact, as we shall see below, equation [12.4] can be easily integrated once to give the dependence of $dV(x)/dx$ and hence of $E(x)$ on x. A second integration to yield the dependence of $V(x)$ on x is more problematical and can only be performed numerically for the general expression. However, we will show that considerable physical insight is attained using an approximation.

12.2   The electric field and voltage at the surface for a given surface charge density

The first integration of equation [12.4] is best performed by using the mathematical identity

$$\frac{d^2 V(x)}{dx^2} = \frac{1}{2}\frac{d}{dV}\left[\frac{dV(x)}{dx}\right]^2.$$

We now integrate both sides of this equation with respect to V and use the fact that integration is the inverse of differentiation so that

$$F(V) = \int_V \frac{d}{dV}[F(V)]dV,$$

where $F(V)$ is any function of V. In this case $F(V) = [dV(x)/dx]^2$ so that we obtain the result

$$\int_V \frac{d}{dV}\left[\frac{dV(x)}{dx}\right]^2 dV(x) = \left[\frac{dV(x)}{dx}\right]^2 = 2\int_{V(x)=0}^{V(x)=V(x)}\left(\frac{d^2V(x)}{dx^2}\right)dV(x).$$

Substitution of the expression for $d^2V(x)/dx^2$ from equation [12.4] into equation [12.5], we obtain

$$\left[\frac{dV(x)}{dx}\right]^2 = -2\sum_i\left(\frac{z_iqn_i}{\varepsilon_R\varepsilon_0}\right)\left[\int_{V(x)=0}^{V(x)=V(x)}\exp\left(\frac{-z_iqV(x)}{kT}\right)dV(x)\right]. \qquad [12.5]$$

To perform the integration of the exponential function over $V(x)$ on the right-hand side of this equation, we use the mathematical identity

$$\int_{V=0}^{V=V}\exp(-\alpha V)dV = \frac{-1}{\alpha}[\exp(-\alpha V) - 1],$$

where in our case $\alpha = z_iq/kT$. Thus, finally, we obtain the required equation for $dV(x)/dx$ or equivalently $E(x)$:

$$E^2(x) = \left[\frac{dV(x)}{dx}\right]^2 = \frac{2kT}{\varepsilon_R\varepsilon_0}\sum_i n_i\left[\exp\left(\frac{-z_iqV(x)}{kT}\right) - 1\right]. \qquad [12.6]$$

To relate the value of the electric field $E(x)$ or the voltage $V(x)$ at the surface, where $x = 0$ so that $E(x) = E(0)$ and $V(x) = V(0)$, to the free charge per unit area σ_S on that surface, we may construct a second Gauss law cylinder with its axis perpendicular to the surface and with one plane face buried in the surface as in Fig. 12.1. We know that no field penetrates the curved sides of the cylinder as the field is parallel to these walls and, if we assume that no field leaves from its plane face within the surface in the negative x-direction (see Section 12.5), we may apply Gauss's law to obtain $D(0) = \sigma_S$ or

$$E(0) = \frac{D(0)}{\varepsilon_R\varepsilon_0} = \frac{\sigma_S}{\varepsilon_R\varepsilon_0}. \qquad [12.7]$$

This is the required relation connecting the surface charge per unit area with the electric field at the surface, $E(0)$. Combining equations [12.7] and [12.6], we obtain the desired relation which connects the surface charge density σ_S with the voltage at the surface $V(0)$:

$$\sigma_S^2 = 2kT\varepsilon_R\varepsilon_0\sum_i n_i\left[\exp\left(\frac{-z_iqV(0)}{kT}\right) - 1\right]. \qquad [12.8]$$

As a simple example of this more general relation, we will assume that the aqueous solution bathing the surface contains a single type of cation with valence $+z$ and a single type on anion with valence $-z$, which each have a concentration per unit volume of n far from the charged surface. Then equation [12.8] becomes

$$\sigma_S^2 = 2nkT\varepsilon_R\varepsilon_0 \left[\exp\left(\frac{-zqV(0)}{kT}\right) + \exp\left(\frac{+zqV(0)}{kT}\right) - 2 \right]. \qquad [12.9]$$

Given a value of σ_S, we can solve this equation to obtain the voltage at the surface $V(0)$. In Table 12.1 we show the prediction for the voltage $V(0)$ at 25 °C for solutions containing only univalent or only divalent ions. The surface is assumed to carry a **negative** charge per unit area of $\sigma_S = -0.1\,\mathrm{C\,m^{-2}}$, which corresponds on average to a single electron charge located within a square area of the surface with sides of magnitude 1.26×10^{-9} m. This is well within the range of the surface charge densities found for the plasma membrane of an animal cell.

The equations we have derived are valid for a positively or negatively charged surface but, essentially because equation [12.6] gives the square of the electric field and thus not its sign, there is a mathematical sign ambiguity which must be resolved if the equations are to make physical sense. In other words, the algebraic sign associated with the expression is not determined by the mathematical equations but must be determined using physical insight. If the surface charge density is negative then the voltage at the surface is negative and vice versa. A positive surface charge density corresponds to an electric field pointing outward from the surface, and a negative surface charge density corresponds to the field pointing towards the surface. Note also that the concentration n_i in the equations above denotes number of ions of a given type per cubic metre and not moles per litre.

Table 12.1   Near a plane which bears a negative surface charge density of $\sigma_S = -0.1\,\mathrm{C\,m^{-2}}$ and is bathed in a aqueous solution of a single univalent–univalent salt ($z = 1$) or a single divalent–divalent salt ($z = 2$), the table gives the voltage near the surface in millivolts for dissolved salt concentrations ranging from 10^{-7} molar to 10^{-1} molar.

Ionic Concentration [moles litre^{-1}]	Surface voltage for $z = 1$ [mV]	Surface voltage for $z = 2$ [mV]
10^{-7}	−434.5	−217.3
10^{-6}	−376.3	−188.1
10^{-5}	−318.1	−159.0
10^{-4}	−259.8	−129.9
10^{-3}	−201.6	−100.8
10^{-2}	−143.2	−71.6
10^{-1}	−83.3	−41.7

The physical reason for the surface voltage reduction with high ionic concentrations is that ions, called **counter-ions**, which carry the opposite sign of charge to that dominant on the surface, are electrically attracted towards the surface, and those ions with a charge of the same sign as the surface charge, called co-ions, are repelled from the surface. These attractive and repulsive forces which tend to separate the ions are combated by the thermal fluctuation forces which tend to randomize the ionic distribution, as we discussed in Section 3.4. The voltage created at the surface by the more abundant counter-ions then cancels some of the voltage at the surface due to the surface charge density. Because the forces due to a given electric field acting upon the ions are twice as great for divalent ions as for univalent ions, the screening effect of the neighbouring ions on the surface voltage is much bigger for divalent ions. From Table 12.1 we see that divalent ions at a concentration of 10^{-5} moles per litre are approximately as effective at reducing the surface voltage as are univalent ions at a concentration 1000 times greater.

12.3   The variation of voltage with distance from the surface

To find an expression for the variation with x of the voltage $V(x)$ we must integrate equation [12.6] with respect to x once again. However, this can only be performed numerically for the general case. If we assume a simple univalent–univalent or divalent–divalent solution, as we did above to obtain equation [12.9], then we can obtain an analytical solution, but the algebraic form of the solution is complicated and provides little insight. The shape of the predicted reduction of $V(x)$ with x for a single dissolved univalent–univalent ($z = 1, z = -1$) salt or a single divalent–divalent ($z = 2, z = -2$) salt is shown in Fig. 12.2 as full curves. We may include both the univalent and the divalent salts at any concentration in the same plot by constructing graphs that are plots of $zqV(x)/kT$ against x/λ like those shown in the figure as full lines. The distance λ is called the **Debye length** and, as we shall see below, it gives a scale to the thickness of the Debye layer for a given concentration of electrolytes in the bathing solution. For a general electrolyte its value is given by

$$\left(\frac{1}{\lambda}\right)^2 = \frac{1}{kT\varepsilon_R\varepsilon_0}\sum_i n_i z_i^2 q^2, \qquad [12.10]$$

which for a single $(+z, -z)$ univalent or divalent salt solution with concentration n becomes

$$\left(\frac{1}{\lambda}\right)^2 = \frac{2nz^2 q^2}{kT\varepsilon_R\varepsilon_0}. \qquad [12.11]$$

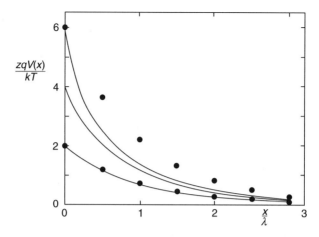

Fig. 12.2   A graph showing the fall with the distance x from the plane charged membrane of the voltage $V(x)$ when the membrane is bathed by an aqueous solution containing only a single univalent–univalent or divalent–divalent salt. The ordinate is $zqV(x)/kT$ and the abscissa is x/λ, where z is the valence of the salt, q is the proton charge, k is the Boltzmann constant, T is the absolute temperature and λ is the Debye length defined in the text.

The three full curves in Fig. 12.2 give the predicted variation of $zqV(x)/kT$ with the ratio x/λ for the three values of $zqV(0)/kT = 2, 4$ and 6. At $20\,°C$ the value of kT/q is 25.3×10^{-3} V so that the three curves correspond to surface voltages of $50.58\,mV$, $101.16\,mV$ and $151.74\,mV$ respectively for univalent ions with $z = 1$. For divalent ions with $z = 2$ the surface voltages must be halved. The plotted curves are for positive surface voltages but the results for negative surface voltages are obtained by simply changing the sign of all the predicted voltages.

In Table 12.2 we give the variation of the Debye length for an aqueous solution of a single univalent–univalent salt with the salt concentration in moles per litre far

Table 12.2   The Debye length λ in units of 10^{-9} m for aqueous solutions of a single univalent–univalent salt with concentrations ranging from 10^{-6} molar to 10^{-1} molar.

Ionic concentration [moles litre^{-1}]	Debye length λ for $z = 1$ [10^{-9} m]
10^{-6}	304.9
10^{-5}	96.9
10^{-4}	30.5
10^{-3}	9.64
10^{-2}	3.05
10^{-1}	0.964

from the charged surface. For a single divalent–divalent salt in aqueous solution, the Debye lengths for the same concentrations are half as great.

An approximation which results in a simple solution which gives considerable insight into the structure of the Debye layer is obtained if we assume that the expression $z_i q V(x)$ is small compared to kT for all values of $V(x)$ encountered. Then we can expand the exponential in equation [12.6] using the mathematical identity

$$\exp(y) = 1 + y + \frac{y^2}{2} + \text{higher powers of } y. \qquad [12.12]$$

Using this expansion and also assuming electrical neutrality within the solution, as discussed in Section 11.1, such that

$$\sum_i n_i q z_i = 0,$$

we obtain from equation [12.6] the simple equation

$$\frac{dV(x)}{dx} = -\frac{V(x)}{\lambda}. \qquad [12.13]$$

The solution of this differential equation is

$$V(x) = V(0) \exp\left(\frac{-x}{\lambda}\right), \qquad [12.14]$$

which predicts a simple exponential fall of $V(x)$ with x from its value $V(0)$ at the charged surface. **The Debye length is seen to measure the thickness of the Debye layer** in that the voltage in the bathing solution falls by a factor $e = \exp(1) = 2.718$ whenever the value of x increases by the Debye length λ. Again this prediction holds for both positive and negative surface voltages simply by changing the signs of $V(0)$ and $V(x)$.

Because the value of kT/q is 25.3 mV at 20 °C, this approximation is only strictly valid for univalent ions when the voltage is much less than 25 mV or for divalent ions when the voltage is much less than about 12 mV. Using this approximation, the variation of $zqV(x)/kT$ with x/λ is shown as filled circles in Fig. 12.2 for values $zqV(0)/kT$ of 2 and 6. It is seen that the approximate expression is in good agreement with the more accurate prediction (given by the full lines) for $zqV(0)/kT = 2$ but becomes less good as the surface voltage rises, as we would expect. Nevertheless this approximation is useful in that the more accurate numerical solutions do predict that the voltage falls with x in a manner that roughly approximates an exponential at higher surface voltages and accurately approximates the exponential at lower surface voltages. The approximation also introduces the Debye length in a particularly simple way which is useful when estimating the thickness of the solution layer next to the charged surface where the effects of its charge are felt.

12.4   The variation of ionic concentration with distance from the charged surface

The distribution of ionic concentration with x results from a competition between the tendency of the electric field created by the charged surface to separate cations and anions and the thermally excited fluctuations that tend to randomize the distribution of ions. As described in Section 12.1, the ratio $n(x)/n(\infty)$ of the concentration of an ion of valence z at a position where the voltage is $V(x)$, to its concentration, remote from the charged surface, where the voltage $V(\infty)$ is zero, is given by the Boltzmann factor

$$\frac{n(x)}{n(\infty)} = \exp\left(\frac{-zqV(x)}{kT}\right).$$

[12.15]

Thus near a positively charged surface where the voltage in the solution is $+116.46\,\mathrm{mV}$, the concentration of univalent cations and anions at a temperature of $20\,^{\circ}\mathrm{C}$ will be reduced by a factor of 100 or increased by a factor of 100 respectively. The concentrations of divalent cations and anions will be reduced and increased by a factor of $10\,000$ in the same voltage. Similarly, a voltage of $58.23\,\mathrm{mV}$ will be responsible for reductions or increases in the concentrations of univalent cations and anions by a factor of 10 and divalent cations and anions by a factor of 100. Thus it is seen that quite moderate voltages created by charged surfaces have a very large effect upon the concentrations of cations and anions within the Debye layer.

In Fig. 12.3 is plotted the logarithm to the base 10 of the concentrations of cations and anions with valence z against the scaled distance x/λ from a charged surface with

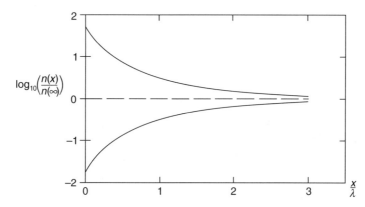

Fig. 12.3   A graph showing the logarithm to the base 10 of the ratio $n(x)/n(\infty)$ plotted against x/λ for a membrane surface voltage $V(0)$ such that $zqV(0)/kT = 4$, where $n(x)$ is the ionic concentration at a distance x from the membrane, $n(\infty)$ is the same ionic concentration remote from the membrane, z is the valence of the single univalent–univalent or divalent–divalent salt dissolved in the bathing solution, q is the proton charge, k is the Boltzmann constant, T is the absolute temperature and λ is the Debye length for the bathing solution.

a surface voltage $V(0)$ such that $zqV(0)/kT = 4$. This is simply obtained from equation [12.15] by using the mathematical identities

$$\log_{10}(y) = \log_{10}(e)\log_e(y) = 0.4343\log_e(y)$$

and

$$\log_e[\exp(y)] = y.$$

Thus we show that

$$\log_{10}\left[\frac{n(x)}{n(\infty)}\right] = 0.4343\log_e\left[\frac{n(x)}{n(\infty)}\right] = 0.4343\left[\frac{-zqV(x)}{kT}\right], \qquad [12.16]$$

where the number in the square bracket on the right is shown as a function of x/λ in Fig. 12.2. We can read off from Fig. 12.3 the concentration ratios for cations and anions of valence z, remembering that

$$\log_{10}(100) = 2, \quad \log_{10}(10) = 1, \quad \log_{10}(1) = 0,$$
$$\log_{10}(0.1) = -1, \quad \log_{10}(0.01) = -2.$$

12.5. How reliable is the simple theory of the Debye layer?

Considering the many simplifying assumptions that are made in the simple theory of the Debye layer, its predictions as to the voltage fall-off with distance from the surface and the resultant concentrations of ions of a given valence are found to be surprisingly accurate (McLaughlin 1989; Bedzyk *et al.* 1990) when tested by experiment.

Among the simplifications made is the fact that no account is taken of the detailed local structure of the solution. Thus although the average concentration of the various ions is related to the local average voltage, no account is taken for example of the fact that surrounding a given cation will be a higher than average concentration of anions and vice versa. Another major assumption is that the bathing fluid is continuous, whereas it may in fact be composed of molecules of considerable size. Thus it would not be reasonable to rely on the theoretical predictions at distances from the surface comparable to hydrated molecular diameters. Also along these lines, as we discussed in Section 5.1, the dielectric constant of water is unlikely to be as high as $\varepsilon_R = 80$ in regions of strong electric fields such as very near the charged surface.

We also made the assumption that the surface charge is smeared uniformly over the surface, whereas in nature it clearly exists as discrete charges distributed over the surface. This in fact makes very little difference except very near the surface. If we assume the charge to exist as discrete charges distributed on the corners of a planar square grid of side d, then at any distance greater than d from the surface, the electric field due to the discrete charges is everywhere within 1% of the field due to a smeared charge of the same surface charge density (Israelachvili 1992). For an assumed surface charge per unit area of $0.1\,C\,m^{-1}$, which corresponds on average to a single proton charge for a square area of side $1.26 \times 10^{-9}\,m$, we can safely ignore the granularity of the charge at all distances greater than $1.26 \times 10^{-9}\,m$ away from the surface.

We derived equation [12.7] using the assumption that no electric displacement $\tilde{D}$ escaped the Gauss law cylinder through its flat end buried in the charged surface. When we considered the parallel plate capacitor in Section 5.4 we made the same assumption, but in that case it was sound because the Gauss law surface was buried in a conductor where the electrostatic field, and hence D, is always zero. For equation [12.7] this is not so and the charged plane could represent the non-conducting outer membrane of a cell. Thus there is no reason why electric displacement should not exit through the buried flat face of the Gauss law cylinder to terminate on free charges of opposite sign within the membrane or in the cell. If there is any leakage of displacement field $\tilde{D}$ inward, it means that the free charge density on the surface gives an upper bound to the outward-pointing displacement field so that equation [12.7] becomes $D(0) \leq \sigma_S$. An alternative but exactly equivalent technique is to extend the length of the Gauss law cylinder through the membrane to include all the charges in the membrane within the cross-section of the cylinder. Now $D(0) = \sigma'_S$ again, but σ'_S becomes the total charge contained per unit area of the surface at any depth and not just that on the outer surface.

A more accurate simulation of the structure of the Debye layer will eventually come from molecular dynamic simulations of the behaviour of aqueous solutions. However, because of the high density of water molecules and their rapid thermally excited motions, such simulations are very expensive in computer time if a reliable equilibrium situation is to be reached. Despite the reservations above, the theory of the Debye layer is certainly accurate enough to predict what is going on to a first approximation and gives us a good simple model of a very complex process.

References

Bedzyk, M. J., Bommarito, G. M., Caffrey, M. and Penner, T. L. (1990) Diffuse-double layer at a membrane–aqueous interface measured with X-ray standing waves. *Science* **248**, 52–56.

Israelachvili, J. (1992) *Intermolecular and Surface Forces*, 2nd edition, Section 12.20. Academic Press, London.

McLaughlin, S. (1989) The electrostatic properties of membranes. *Annual Review of Biophysics and Biophysical Chemistry* **18**, 113–136.

13.   The behaviour of ions in narrow pores

Essentially because the ions within a solution passing rapidly through a narrow pore may not be in equilibrium with each other or with the charges on the walls of the pore, the behaviour of ions passing through a narrow pore is more difficult to predict than the behaviour of the same ions in bulk solution. In addition, the electrostatic self-energy of an ion in a narrow pore is very different from that of the same ion in bulk solution. A simple model of this enhanced self-energy is presented, and stronger electrostatic inter-ionic interactions within the pore are shown to be a consequence. The mechanisms that may lead to partial ordering of the electric dipole moments of water molecules in the pore are discussed, and the consequences of such order on the prediction of the ionic current through ion channels are described.

13.1   Ion channels in biology

Many biological processes are controlled by the passage of ions across a membrane, through membrane ion channels, which are proteins that span the membrane and are currently thought to provide narrow aqueous pores that convey the ions across the membrane. The most obvious example is the transmission of the action potential along an axon, but there are many others. In Section 5.8 we described how the large rise in the electrostatic self-energy of a small ion moving from an aqueous environment with a high relative dielectric constant to the lipid bilayer with a low relative dielectric constant provides a formidable barrier to the passage of small ions across a lipid bilayer membrane. Thus one important function of the ion channels is to provide a low-energy path across the membrane for small ions. Other important functions are selectivity so that only certain types of ion are allowed to pass through the channel, and switching such that the channel opens and shuts in response to the voltage across the membrane or to the binding of specific ligands to the trans-membrane channel protein.

   Despite a great deal of experimental information (Hille 1992) about the properties of particular types of ion channel and even the determination of the molecular structures of a few ion channels, the molecular mechanisms which lead to their selectivity and gating are as yet poorly understood. One reason for this may be the fact that during the time the ions occupy the channel when in transit, which may typically be as short as 1 μs, they may not be in equilibrium with their surroundings. A bulk ionic solution is in thermodynamic equilibrium and as a result it is highly homogeneous in structure and any local deviations from equilibrium are quickly corrected, as discussed in Section 11.1. The composition of the solution passing through a narrow pore may have been determined by the conditions that prevailed at the pore ends and need not be in equilibrium with the material of the pore walls. Thus equilibrium properties, such as pH or the electrochemical potential of one type of ion μ_i, have little meaning when applied to the solution passing through a narrow channel. Other factors that complicate theoretical predictions of the properties of ions within narrow pores are the rise in their electrostatic self-energy and the fact that electrostatic interactions within the pore can be very different from those in bulk solution. Yet another complication is the possibility that the water within a narrow pore may be partially ordered and thus exhibit different behaviour from bulk aqueous solution. In this chapter we will briefly consider some of the possible differences between the properties of ions in bulk solution and those of ions in narrow water-filled pores.

13.2   The electrostatic self-energy of an ion in a narrow water-filled pore

In Sections 5.5 and 5.8 we discussed the rise in the electrostatic self-energy of an ion when it is moved from an electrically polarizable environment with a high relative dielectric constant to an environment where the electrical polarizability and thus the relative dielectric constant are small. The physical mechanism operating is made clear if we compare the situations of a single cation with charge q, firstly in an aqueous solution with high electrical polarizability, and secondly within a lipid bilayer where the electrical polarizability is low. In the aqueous solution the electrical dipole moments of the surrounding water molecules tend to point their negative ends toward the cation, as described in Section 11.1, which lowers the voltage V_S at the surface of the ion and thus its electrostatic self-energy $\frac{1}{2}qV_S$. In the lipid there are only very weak induced dipoles so that V_S and the self-energy remain high.

An ion in a narrow water-filled pore spanning a lipid bilayer is in an intermediate situation in that it is partially hydrated and has an electrostatic self-energy higher than that of an ion in bulk aqueous solution but lower than that of an ion in bulk lipid (Parsegian 1969). A quantitative physical explanation of the rise in the self-energy when an ion leaves the solution and enters the pore is as follows. Let us consider a single cation on the axis of a long but narrow cylindrical pore filled with aqueous solution with relative dielectric constant ε_{R1}. Let the pore penetrate a block of

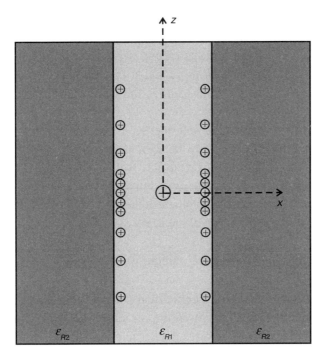

Fig. 13.1 A cylindrical pore with its axis along the z-axis filled with an aqueous solution of relative dielectric constant ε_{R1} which penetrates a block of material with relative dielectric constant ε_{R2}. A cation is shown on the axis of the pore and the distribution of the positive charge induced on the surface of the pore, if $\varepsilon_{R1} > \varepsilon_{R2}$, is indicated.

solid dielectric with relative dielectric constant ε_{R2} such that $\varepsilon_{R1} > \varepsilon_{R2}$, as sketched in Fig. 13.1.

Within the solution surrounding the charge there is no polarization charge density induced by the electric field of the ion essentially because $\mathrm{div}(\tilde{P}_1) = 0$, where $\tilde{P}_1$ is the electric dipole moment per unit volume induced in the solution by the electric field of the ion, as described in Section 5.2. Physically this is because any spherical shell surrounding the charge contains as much positive charge from the positive ends of the induced dipoles as negative charge from their negative ends. However, at the surface of the solution where it meets the pore wall there is induced a surface charge per unit area equal to the outward-pointing normal component of the solution polarization P_{N1}, as discussed in Section 4.3. This is due to the uncompensated positive ends of the induced dipoles that protrude normally outward from the surface of the fluid cylinder. At this interface there is also induced a surface charge per unit area of $-P_{N2}$ from the uncompensated negative ends of the dipoles induced by the electric field of the ion in the solid dielectric, which point along the normal into the solid dielectric surface. Thus on the surface between the solution and the solid dielectric there is a net surface charge per unit area of $P_{N1} - P_{N2}$. Because $\varepsilon_{R1} > \varepsilon_{R2}$, at the pore wall we have $P_{N1} > P_{N2}$ so that the surface charge induced by a cation is positive, as sketched in Fig. 13.1. Following this same argument for an anion, it is seen that in general the induced surface charge always has the same sign as the charge on the ion that induces it, provided that $\varepsilon_{R1} > \varepsilon_{R2}$.

The change in the electrostatic self-energy of the ion when it leaves the bulk solution and enters the pore may now be seen as due to the electrostatic interaction between the induced surface charge on the pore wall and the charge of the original ion. As the induced surface charge has the same sign as the charge on the ion, the change in self-energy is always positive. If the voltage at the ion generated by this surface charge is V_S, the change in the self-energy of the ion U_S is given by

$$U_S(\text{pore}) - U_S(\text{solution}) = +\frac{1}{2}qV_S. \qquad [13.1]$$

For a univalent ion on the axis of a cylindrical pore of radius a, filled with liquid of relative dielectric constant 80 representing an aqueous solution, which penetrates a block of solid dielectric with relative dielectric constant of 2 representing the lipid bilayer, the value of V_S is $2.44 \times 10^{-10}/a$ volts, so that from equation [13.1] we obtain the rise in the self-energy as

$$U_S(\text{pore}) - U_S(\text{solution}) = \frac{1.95 \times 10^{-29}}{a} \text{ J}. \qquad [13.2]$$

The magnitude of the voltage generated by the pore wall surface charge at the ion, V_S, was calculated in this cylindrical geometry using Bessel functions. The method has been described clearly elsewhere (Parsegian 1969; Smythe 1989) but is too complex for inclusion here.

For a pore of radius $a = 10^{-9}$ m, the rise in self-energy for a univalent ion is $\Delta U_S = 1.95 \times 10^{-20}$ J. We may apply the Boltzmann ratio $P(\Delta U_S)/P(0) = \exp(-\Delta U_S/kT)$, discussed in Appendix 2, for the ratio of the probabilities of finding the ion at positions where its energy differs by ΔU_S. By these means we find that at a temperature of 25 °C, the ion is 114 times less likely to be found in the pore than in the solution. For a divalent ion, the rise in self-energy ΔU_S is four times as large, because both q and V_S in equation [13.1] are doubled in magnitude, and the ratio of the Boltzmann probabilities becomes about 172 million for the model pore considered above.

Calculations such as these which treat the water in the pore as continuous, rather than granular with a molecular diameter comparable with that of the pore, and which assume that the relative dielectric constant of water remains fixed no matter how high the electric field, are unlikely to be accurate, as mentioned in Section 5.8. However, they do point to the fact that the rise in electrostatic self-energy is a major obstacle to the passage of small ions through narrow pores and that a major function of the membrane spanning proteins that form biological ion channels is to lower the self-energy so that a substantial ion flux through them is possible.

13.3   Enhanced electrostatic interactions within narrow water-filled pores

The charge that is induced at the surface of a narrow water-filled pore by an ion within the pore, which is responsible for the rise in the self-energy of the ion within the pore that we discussed above, is also responsible for greatly enhanced electrostatic interactions between ions in such narrow pores (Edmonds 1994). The electric field generated by an ion in a pore may be thought of as having two origins. The first is the field directly generated by the ion, which has a magnitude equal to that of the same ion in a bulk solution of the fluid that fills the pore. The second contribution to the electric field at an ion is that generated by the surface charge induced on the surface of the pore as described above. The rigorous justification of the equivalence of these two super-imposed electric fields to the total field within a pore comes from the analytical solution of the electrostatics of a charge on the axis of a long cylindrical pore using Bessel functions, which is referred to above (Smythe 1989). The second contribution, which is caused by charge induced on the pore wall and so is missing in bulk solution, is responsible for the greatly enhanced electric fields and voltages generated by a single ion within a narrow pore.

Figure 13.2a shows the contours of the voltage in millivolts surrounding a univalent cation on the axis of a long cylindrical pore of radius $a = 10^{-9}$ m filled with fluid with relative dielectric constant $\varepsilon_{R1} = 80$ which is surrounded by a solid block with relative dielectric constant $\varepsilon_{R2} = 2$. The vertical axis in the figure represents the distance along the cylindrical axis z and the horizontal axis represents the radial distance r from the cylindrical axis. These distances are scaled such that the vertical axis shows z/a and the horizontal axis shows r/a, where a is the radius of the pore. The complete three-dimensional picture within the pore may be visualized by rotation of this figure about the z-axis and reflection in the (x, y) plane. Figure 13.2b shows the spherical contours of the voltage in millivolts that would be generated by the same ion at these distances in a bulk solution with relative dielectric constant ε_{R1}. It can be seen that the voltages generated in the pore are much bigger and of much longer range than those that are generated in a bulk solution of the same fluid that fills the pore.

Before considering the consequences of the enhanced fields in the pore, we will consider further the physical origin of the enhancement of the fields generated within

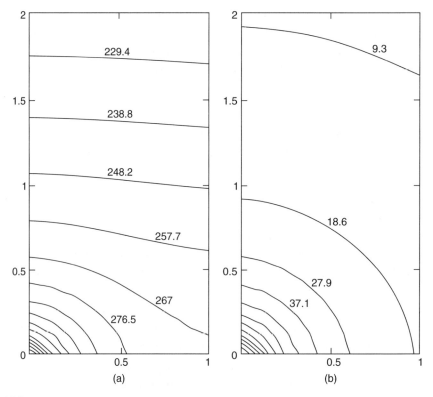

Fig. 13.2   (a) A contour map of the voltage in millivolts surrounding a univalent cation on the axis of a cylindrical water-filled pore of radius a which has the z-axis as its axis. The ordinate gives the scaled distance z/a along the z-axis, where z is the axial distance from the centre of the cation, and the abscissa gives the scaled distance x/a, where x is the perpendicular distance away from the axis of the pore. The relative dielectric constants assumed for the solution in the pore and for the material surrounding the pore are $\varepsilon_{R1} = 80$ and $\varepsilon_{R2} = 2$. (b) The same voltage contours for the univalent ion in bulk solution against the same scaled distances from the ion when the relative dielectric constant for the solution is $\varepsilon_{R1} = 80$. In three dimensions the contours in this case are spherical.

the narrow pore. If we construct a spherical Gauss law surface, concentric with the ion and close to it, we deduce, using Gauss's law, that the total flux of the electric field $\tilde{E}$ through this surface is $q/\varepsilon_{R1}\varepsilon_0$, where q is the charge on the ion. This is the same flux as would be created by the same charge in bulk solution and no charges outside the sphere can change this flux. The enhancement of the field within the pore is due to a redistribution of the electric field throughout space and not to an enhancement of the total flux of the electric field leaving the ion. When situated in bulk solution, the field of the ion is radial and equally distributed in every direction. This can be seen in Fig. 13.2b when it is remembered that the field direction is everywhere perpendicular to the contours of the voltage which in this case are spherical and centred on the ion. In Fig. 13.2a it can similarly be seen that the direction of the field, which is radial close

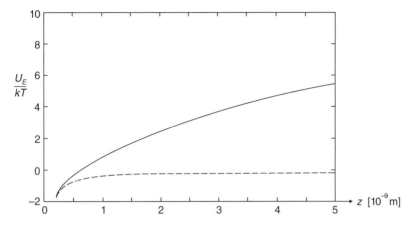

Fig. 13.3   A graph showing as a full line the excess energy U_E (which includes their self-energies) for a univalent cation and anion pair on the axis of a water-filled pore of radius 10^{-9} m plotted against the distance between the ions in units of 10^{-9} m. The ordinate shows the scaled energy U_E/kT, where k is the Boltzmann constant and T is the absolute temperature. The zero of excess energy is taken as the energy of the pair, including the self-energies, when they are widely separated in bulk solution. The dashed line gives the same scaled energy plotted against separation for the same pair of ions in bulk solution. The solution and the material surrounding the pore are assumed to have relative dielectric constants of 80 and 2.

to the ion, changes direction so that much of it is directed parallel to the axis of the pore and very little penetrates through the pore walls. It is this channelling of the field along the pore where the relative dielectric constant is greatest that leads to enhanced electric fields within the pore and hence to enhanced electrostatic interactions within the pore.

In Fig. 13.3 we show as a full line the calculated total electrostatic interaction energy U_E of a univalent cation and a univalent anion both on the axis of a narrow water-filled pore, plotted against their separation z along the axis of the pore measured in nanometres. The chosen characteristics of the pore are the same as those for Fig. 13.2, namely, $a = 10^{-9}$ m, $\varepsilon_{R1} = 80$ and $\varepsilon_{R2} = 2$. In the figure the electrostatic energy is shown as the ratio U_E/kT, where $kT = 4.12 \times 10^{-21}$ J is a typical thermal energy at 25 °C, and the zero of energy represents the state when the ions are widely separated and in the bulk solution. When the ions are both in the pore but widely separated, U_E is the sum of the two self-energies given by equation [11.2] so that $U_E/kT = 9.48$ which is just below the top of the Fig. 13.3. As the ions approach each other, the cation interacts directly with the anion but also with the negative charge induced at the pore wall near the anion, and the electrostatic energy is reduced. When the ions are close together on the pore axis the charges of opposite sign induced at the pore wall by the two ions tend to cancel to zero and the electrostatic energy approaches that between the same two ions in bulk solution. For comparison, the electrostatic energy of the two ions in bulk solution is plotted against their separation in Fig. 13.3 as the lower broken line curve. By comparison between the full and broken curves it can be seen that the electrostatic interaction energy of the ion pair in the pore is much greater in amplitude and much longer in range that the identical interaction in bulk solution.

One consequence of the enhanced electrostatic interaction within a water-filled pore could be the binding of a cation and an anion to form a salt molecule rather than remaining as separate ions (Edmonds 1994). In bulk solution the degree of dissociation of a salt into separate ions is determined by the difference in the Gibbs free energy between the bound and separated states. The large rise in the enthalpy when the ions are separated within a narrow pore, which we discussed above, will tend to favour the combined state. Thus ion pairs that are fully dissociated in bulk solution may be partially associated while in the pore. Voltage gradients that drive charged ions along the pore will not operate on the neutral salt molecule and the diffusion of the salt through the pore will convey no charge. Both these effects will reduce the measured current flow through the pore.

13.4   Interactions between ions and ionizable residues in the pore wall

In Section 11.3 we considered the ionization of acid and basic residues that are responsible for charged groups in proteins. We also showed in Section 11.4 that an externally applied local positive voltage increases the probability of finding an acid residue in its negatively charged state by reducing the concentration of hydrated protons near the residue. These effects can lead to co-operative interactions between cations and acid residue sites or between anions and basic residue sites. Let us consider as an example an acid site embedded in the wall of a water-filled narrow pore. The site on the pore wall has a much less electrically polarizable environment than would be the case if it were exposed on a surface bathed by the bulk solution, so that the probability of finding it in a charged state is much reduced, as described in Section 11.5. However, the voltage at the site created by a cation passing through the pore will increase the probability that the acid wall site is in its negatively charged state. The negative voltage created by the negatively charged acid site will in turn lower the electrostatic energy of the cation within the pore. Thus the cation in transit can create an electrical environment which aids its own passage through the pore. Such considerations lead to the view that the charge structure of an ion channel is unlikely to be static but will probably respond transiently to the passage of ions through the pore (Edmonds 1989). Co-operative interactions such as these may lead to a correlation between the charged state of a pore wall and the prevailing ion flow through the pore. An expanded discussion of some of these topics and a comparison with experiments may be found in Edmonds (1998).

We will discuss in Chapter 15 a simple electrostatic model of a proton/ion or an ion/ion coport or counterport that relies upon the dynamic interaction of ions within a pore and residues buried in the pore wall. Coports and counterports transfer a particular type of ion across the cell membrane against the prevailing electrochemical gradient for that type of ion, using the drop in the electrochemical potential of another type of ion simultaneously or sequentially crossing the membrane. Such models require no mechanical motion to function other than the motion of protons that change the charged state of the residue and the passage of the ions through the pore.

13.5   The possible ordering of the water structure within narrow pores

If the directions of the electric dipole moments of the water molecules within a narrow pore prove to have an appreciable ferroelectric order, then the treatment of ionic transport through the pores will have to change markedly. Whether such order can exist in liquid water remains controversial today, although there does exist experimental evidence that the presence of impurities or the proximity of a suitable substrate can induce such order in normal hexagonal ice (Bramwell 1999).

There are two distinct reasons why the water in narrow pores may have a structure different from that of bulk water. The first is the change in water structure near a hydrophobic or hydrophilic surface. A hydrophobic surface may be defined as one that incorporates few charged sites or hydrogen bond donor or acceptor sites. The water molecules near a plane hydrophobic surface have an energy higher than water molecules within the solution because they are unable to participate in as many hydrogen bonds as they can in bulk solution, as discussed in Section 11.1. To minimize the resultant rise in energy it is thought that each water molecules next to a plane hydrophobic surface is likely to orient one of its four possible tetrahedral hydrogen-bonding orbitals normally to the surface which allows the other three tetrahedral orbitals to participate in hydrogen bonding with neighbouring water molecules. Any ordering will be opposed by the resultant fall in entropy (see Appendix 2) of the partially ordered water molecules near the surface. There is some evidence, both theoretical with molecular dynamic simulations (Lee and McCammon 1984; Zhu and Robinson 1991), and experimental in vibrational spectroscopy (Du *et al.* 1994) and X-ray scattering (Toney *et al.* 1994), which seems to support some ordering of the water structure for the first two or three layers of water next to the hydrophobic surface. Near a hydrophilic surface, the energy loss caused by the reduced number of water-to-water hydrogen bonds may be compensated by hydrogen bonds from adjacent water molecules to the exposed charged and hydrogen bonding groups of the surface, and this could also lead to some ordering of the water structure. Any ordering that does occur will of course be dynamic, but the hydrogen bonding pattern is thought to persist for much longer than the typical half-life of a hydrogen bond in water of about 10^{-11} s. In a very narrow pore with a diameter of 10^{-9} m, the vast majority of water molecules will be nearest or next nearest neighbours of the pore wall and thus could be partially ordered.

The second mechanism that may lead to the ordering of water is specific to water within a narrow pore, and has its origins in the electrical interaction between the strong electric dipole moments of the water molecules. The electrical interaction energy of two dipoles is low when they are end to end and pointing in the same direction because the electric field of the first dipole $\tilde{E}_1$ points along the direction of the dipole moment $\tilde{p}_2$ of the second dipole and the energy becomes $-\tilde{p}_2\tilde{E}_1$ as in equation [4.5]. If the two dipoles remain pointing in the same direction but are now placed side by side, instead of end to end, the energy rises because the field of the first dipole $\tilde{E}_1'$ is now pointing oppositely to the dipole moment $\tilde{p}_2$ of the second (see Fig. 4.1a) and the energy becomes $+\tilde{p}_2\tilde{E}_1'$. An elongated and narrow water structure such as the water confined within a narrow pore maximizes the number of end-to-end dipole pairs and minimizes the number of side-by-side pairs and the array can thus lower its electric energy if as many dipoles as possible point in the same direction approximately along the pore axis. The effect is co-operative in that the greater the number of dipoles aligned approximately along the pore axis, the greater the electric field in this direction and the greater the energy drop of other dipoles turning to point in this direction.

Any dipolar alignment that reduced the number of water-to-water hydrogen bonds would raise the energy of the water array, but the electric dipole moment of a water molecule can point in many directions while maintaining the full tetrahedral hydrogen bonding network. Thus the lowest-energy configuration of the water molecules will be that which maintains the hydrogen bonding network intact but within which the total dipole moment along the axis is maximized. Both lattice sum calculations (Edmonds 1984) and molecular dynamic simulations applied to the gramicidin A ion-channel (Mackay et al. 1984; Kim et al. 1985; Chiu et al. 1989) support the possibility of this type of dipolar order in a narrow water-filled pore. Such an aligned array will have a macroscopic electric dipole moment which is the sum of the time-averaged molecular dipole moments of the individual water molecules in a given direction, and this macroscopic dipole moment may interact with the strong electric field across the membrane of a cell and participate in the voltage-activated switching of the channel (Edmonds 1984).

If for either of these reasons the water structure in a narrow pore is partially ordered, it will change markedly the diffusion of ions through the pore. The structure of the hydrated zeolites may be thought of as a series of approximately spherical cavities connected to narrow cylindrical pores through narrow necks. This structure is filled with water and small cations have a large mobility through the crystal. The steric activation barrier to cation migration only starts to rise rapidly when the **unhydrated diameter** of the ions becomes as big as the diameters of the narrow necks (Barrer and Rees 1960; Barrer *et al.* 1963). This has led to a model in which the water molecules rotate but do not translate and only the bare cations translate through the crystal being at all times hydrated by the quasi-static local water array. The ions may be thought to tunnel between the centres of adjacent water rings. If the water in narrow membrane channels has a structure like that thought to exist in the hydrated zeolites, the theoretical treatment of the passage of ions through them will need to be completely revised (Edmonds 1984).

References

Barrer, R. M., Bartholomew, R. F. and Rees, L. V. C. (1963) Self and exchange diffusion of ions in chabazites. *Journal of the Physics and Chemistry of Solids* **24**, 51–62.

Barrer, R. M. and Rees, L. V. C. (1960) Self diffusion of alkali metal ions in analcite. *Transactions of the Faraday Society* **56**, 709–721.

Bramwell, S. T. (1999) Ferroelectric ice. *Nature* **397**, 212–213.

Chiu, S., Subramaniam, S., Jakobsson, E. and McCammon, J. A. (1989) Water and polypeptide conformations in the gramicidin channel. *Biophysics Journal* **56**, 253–261.

Du, Q., Freysz, E. and Shen, Y. R. (1994) Surface vibrational spectroscopic studies of hydrogen bonding and hydrophobicity. *Science* **264**, 826–828.

Edmonds, D. T. (1984) The ordered water model of membrane ion channels, In: *Biological Membranes*, Ed. D. Chapman, Volume 5, Chapter 10. Academic Press, London.

Edmonds, D. T. (1989) A kinetic role for ionizable sites in membrane ion channels. *European Biophysics Journal* **17**, 113–119.

Edmonds, D. T. (1994) Enhanced ionic interactions in narrow pores and ion pair formation. *European Biophysics Journal* **23**, 133–138.

Edmonds, D. T. (1998) The behaviour of ions in narrow water-filled pores. *Bioscience Reports* **18**, 313–327.

Hille, B. (1992) *Ionic Channels in Excitable Membranes*, 2nd edition. Sinauer Associates, Sunderland, MA.

Kim, K. S., Nguyen, H. L., Swaminathan, P. K. and Clementi, E. (1985) Na^+ and K^+ ion transport through a solvated gramicidin A transmembrane channel: Molecular dynamics studies using parallel processors. *Journal of Physical Chemistry* **89**, 2870–2876.

Lee, C. Y. and McCammon, J. A. (1984) The structure of liquid water at an extended hydrophobic surface. *Journal of Chemical Physics* **80**, 4448–4454.

Mackay, D. H. J., Berens, P. H. and Wilson, K. R. (1984) Structure and dynamics of ion transport through gramicidin A. *Biophysical Journal* **46**, 229–248.

Parsegian, A. (1969) Energy of an ion crossing a low dielectric membrane: Solutions of four relevant problems. *Nature* **221**, 844–846.

Smythe, W. R. (1989) *Static and Dynamic Electricity*, 3rd edition, Section 5.351. Hemisphere New York.

Toney, M. F., Howard, J. N., Richer, J., Borges, G. L., Gordon, J. G., Melroy, O. R., Wiesler, D. G., Yee, D. and Sorensen, L. B. (1994) Voltage-dependent ordering of water molecules at an electrode–electrolyte surface. *Nature* **368**, 444–446.

Zhu, S. B. and Robinson, G. W. (1991) Structure and dynamics of liquid water between plates. *Journal of Chemical Physics* **94**, 1403–1410.

14. Possible mechanisms for a magnetic animal compass

One important application of magnetism in biology is the magnetic compass, used by many animals for navigation while migrating. The form of the magnetic field of the Earth and some general facts about the animal compass are described. At this time the transducer used in the animal compass is not known but magnetic induction, the presence of magnetite crystals and the lifetime of free radicals are discussed as possible mechanisms for the compass.

14.1. The magnetic field of the Earth

To a surprisingly good first approximation, the magnetic field of the Earth may be modelled as if it arose from a south-pointing magnetic dipole at the centre of the Earth, as sketched in Fig. 14.1. The magnetic dipole moment is thought to have its origin in currents circulating within the molten core of the Earth. The dipole does not point along the axis of rotation of the Earth but is currently inclined to it at an angle of about 11° so that the magnetic north pole is north of Victoria Island, Canada, and the magnetic south pole is on the edge of Antarctica, due south of Tasmania. An even better model of the Earth's field is obtained by displacing the dipole about 500 km from the centre of the Earth towards the western Pacific, but for simplicity we will model the field as due to a magnetic dipole at the centre of a spherical Earth.

The vertical B_V and horizontal B_H components of the magnetic field on the surface of a sphere of radius R, due to a magnetic dipole moment m at its centre, at a position specified by the angle θ to the dipole axis, as shown in Fig. 14.1, are given by Equations [6.4] as

$$B_V = \frac{2\mu_0 m \cos(\theta)}{4\pi R^3},$$

$$B_H = \frac{\mu_0 m \sin(\theta)}{4\pi R^3}.$$

[14.1]

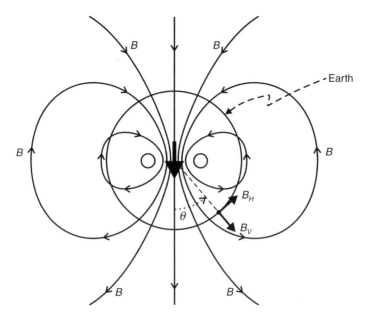

Fig. 14.1   The magnetic field of the Earth shown as due to a single downward-pointing magnetic dipole located at the centre of a spherical Earth. At positions on the Earth's surface defined in the diagram by the angle θ, the vertical and horizontal components of the Earth's magnetic field are shown as B_V and B_H respectively.

The angle of inclination or dip, θ_D, which is the angle between the direction of the field and the horizontal plane, taken as the tangent plane to the spherical surface, is then given by

$$\tan(\theta_D) = \frac{B_V}{B_H} = 2 \cot(\theta). \tag{14.2}$$

Thus in the model, the amplitude of the field is twice as great at the magnetic poles as it is on the magnetic equator, and the field is vertical at the poles ($\theta_D = 90°$) and horizontal ($\theta_D = 0$) at the equator. The experimentally measured magnetic field of the Earth does show a reasonably smooth variation of dip angle as predicted by the model in equation [14.2]. The total amplitude of the field is less well predicted by the model and is measured as 61 microtesla (μT) at the north magnetic pole and 67 μT at the south magnetic pole, with 30 μT as a typical amplitude around the magnetic equator. However, the minimum value near the middle of the South Atlantic is only 23 μT.

There are many reasons why the real field is not quite so regular in its variation as the model predicts. One is the offset of the position of the generating dipole from the Earth's centre mentioned above, and another is the presence of magnetic rocks close to the surface which can cause local anomalies. Nevertheless, a model field which is vertical with an amplitude of about $60 \, \mu T$ at the magnetic poles and is horizontal with an amplitude of about $30 \, \mu T$ around the magnetic equator is sufficiently accurate to judge the plausibility of the mechanisms that have been proposed for the animal magnetic compass.

14.2   The animal compass

It is now well established that many animals have a magnetic compass (Wiltschko and Wiltschko 1995) although the physical mechanism by which it operates is as yet unknown. The major problem when postulating possible mechanisms is the smallness of the Earth's magnetic field and thus the difficulty of obtaining a reliable compass reading in the presence of thermal fluctuations. The kinetic energy stored in the mechanical motions of the molecules in a cubic metre of water is about $2.07 \times 10^8 \, J$ at $25\,°C$, whereas the magnetic field energy $\tilde{B} \cdot \tilde{H}/2$ stored by a magnetic field of $B = 50 \, \mu T$ which occupies the same unit volume is only $9.95 \times 10^{-4} \, J$, which is 2.08×10^{11} times smaller. Of course, there are many examples of **trigger mechanisms** in biology where a small initiating effect results in a robust response. One example is in the eye, where the arrival of four or five photons can trigger a cascade that results in the transfer of many millions of ions across a membrane. The magnetic compass mechanism may rely upon similar amplification.

When reading the literature it is important to realize that an animal may have access to several compasses (Wiltschko and Wiltschko 1995). During cloudless days, many animals use a solar compass which requires a measurement of the position of the Sun and the reading of an internal biological clock to judge direction. During the night a moon compass, used like the Sun compass, is sometimes employed and a stellar compass based upon the apparent circulation of the stars around the geographic pole positions may also be used. Thus experiments designed to distinguish the magnetic compass from other possible compasses must be carefully designed. However there is some evidence (Weindler et al. 1996) that the magnetic compass may be fundamental in calibrating the other compasses in that the technique for using the magnetic compass is passed genetically to the young, without the need for instruction from or indeed any direct contact with, the adult animal.

A second fact of fundamental importance is that the animal may not use the direction of the horizontal component of the Earth's magnetic field as a direction indicator, as we do when using the horizontal pivoted compass needle. In fact many birds and some newts measure the inclination of the field (Wiltschko and Wiltschko 1995) and judge the north–south plane as that within which the direction of the magnetic field makes the smallest angle with the vertical. Such a north–south plane detector clearly works in the northern and in the southern magnetic hemispheres. The compass of such animals detects the direction of the magnetic field axis but not the sign of the field so that artificially reversing the direction of the magnetic field along the same axis has no effect upon their ability to orient correctly.

Having a compass is of little value in determining the direction to a given goal unless the current position relative to the goal is known. In other words, a compass is of little use without a map. There has been much speculation as to whether animals may use the variation of two properties of the Earth's magnetic field, such as the dip angle and total magnetic field amplitude or the dip angle and the gradient of the dip angle, to establish a positional map. The present position is not clear but it is obvious that the accuracy with which the direction and magnitude of the magnetic field must be measured is much higher to establish a map than to simply determine an approximate compass direction. Local magnetic anomalies must also be understood to establish a map. The evidence for a magnetic animal compass is now overwhelming but the possibility that animals have established a reliable magnetic map is highly controversial.

14.3   Magnetic induction

As discussed in Section 9.1, it is possible in principle to measure a magnetic field by measuring the emf induced along a conductor moving in the magnetic field. In Fig. 14.2a is shown a vertical linear conductor AB of length L aligned along the z-axis and moving at a velocity v in the x-direction perpendicular to a horizontal magnetic field of amplitude B_H pointing in the y-direction. We will now calculate the voltage induced between the ends of the conductor. The simplest method is to imagine charges q within the conductor being moved with the conductor perpendicular to the prevailing magnetic field. They will experience a force qvB_H, as shown in equation [10.2], which in this case is along the rod and in the positive z-direction. Positive charges will be driven upward along the rod in the positive z-direction and negative charges will be driven downward along the rod in the negative z-direction. Positive charges will accumulate at the top of the rod and negative charges at the bottom, and these accumulating charges will create an internal electric field E_{int} directed downward along the rod. The charge motion will cease when the upward force qvB_H on a charge q due to the motion in the magnetic field is cancelled by the force qE_{int} exerted by the internal electric field so that $E_{int} = vB_H$. The linear voltage drop between the top and bottom of the rod, which is sketched in Fig. 14.2a, is thus given by

$$V(\text{top}) - V(\text{bottom}) = E_{int}L = vB_HL. \tag{14.3}$$

If we take values of $L = 0.1\,\text{m}$, $v = 10\,\text{m s}^{-1}$ and $B_H = 30\,\mu\text{T}$ as typical, then the voltage across the ends of the moving conductor becomes $30\,\mu\text{V}$ which is very small.

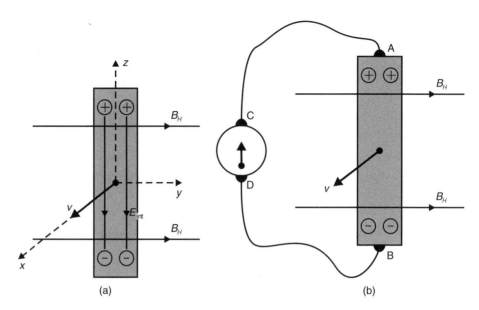

Fig. 14.2 (a) A linear conductor, pointing along the z-axis, is shown shaded and is moving with velocity v in the x-direction in a region where there exists a magnetic field B_H pointing along the y-axis. Faraday's law predicts that an electric field is induced which will drive positive mobile charges in the conductor along the positive z-direction and negative mobile charges in the opposite direction. Equilibrium is established when the accumulation of positive and negative charges at the upper and lower ends of the conductor creates a downward directed electric field E_{int} which cancels the electric field induced by the motion of the conductor through the magnetic field. (b) In an attempt to measure the voltage induced across the ends of the moving conductor shown in (a), leads are attached to these ends and connected to a voltmeter. In fact if the conductor, the leads and the voltmeter are fixed relative to each other, no voltage will register on the voltmeter when the whole apparatus moves with velocity v in the x-direction cutting the field lines of a magnetic field directed along the y-direction. This is because a voltage will be induced across the leads by the motion which will exactly cancel that induced across the original conductor.

To detect the direction of the maximum magnetic field, as performed by birds, would require the bird to fly in accurately known directions at accurately known speeds and detect small differences, which makes this method impractical for this purpose.

Besides the smallness of the induced voltage, a second problem arises if we attempt to measure the voltage difference with a single voltage detector. In Fig. 14.2b the conductor AB is shown as before, but also shown are the extra conductors AC and BD required if the voltage is measured by a single detector located between C and D. The whole apparatus, including conductors and detector, is assumed to move in the x-direction with a velocity v. In fact, with this circuit the voltage induced across the detector by the motion is negligibly small so that this circuit will not measure magnetic fields. This can be understood in two ways. In the first we apply the Faraday law as in equation [9.1]. The circuit $CABD$ has a constant area and if it moves in a region where the magnetic field does not vary, the flux of the magnetic field through the circuit is constant. Thus Faraday's law predicts that the emf induced around the complete circuit is zero. In the second method we employ our original argument and calculate the voltage difference induced between the ends of a conductor moving in a magnetic field. We find that the voltage drop between A and B induced by the movement through the magnetic field is cancelled by the combined voltage drops induced along the conductors CA and BD by the same motion, so that again there is no voltage across the detector.

One class of animals – elasmobranchs, such as sharks, skates and rays – has an electric field detector (Kalmijn 1982; Paulin 1995) that has the sensitivity to measure small magnetic fields by induction; these animals have an ingenious method of measuring the induced voltage caused by motion of a conductor in a magnetic field using a single voltage detector. Their electric field detectors, which are called ampullae of Lorenzini, consist of two mucus-filled pores AB and CD situated along a straight line with a voltage detector between B and C. To describe their function we choose two such pores disposed along a vertical axis when the animal swims in a horizontal plane as sketched in Fig. 14.3. The mucus has an electrical conductivity comparable to sea water but the walls of the pores and the voltage detector have a high

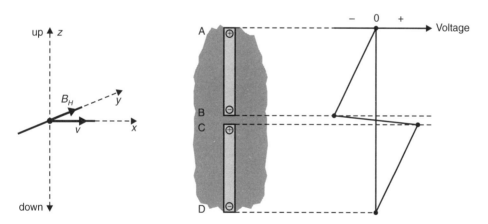

Fig. 14.3 On the left is shown a set of Cartesian axes which define the directions of the motion v along the x-axis with the magnetic field B_H directed along the y-axis. The vertical axis is taken to be the z-axis. On the right is shown a idealized sketch of the ampullae of Lorenzini possessed by many elasmobranch fishes such as skates and rays. The structure consists of a pair of opposed vertical pores with voltage detectors located between their inner ends. The walls of the pores have a high resistance but they are filled with mucus with a conductivity comparable to sea water. Essentially because the outer ends of the pores are electrically connected together by the conducting sea water that surrounds the fish, the total voltage that is induced along the whole length of the two pores appears across the internal voltage detectors as shown by the graph on the extreme right of the diagram.

resistance. If a vertical upward-pointing electric field E_V is present, this will induce charge movement in the two conducting pores, leading to positive charge migrating towards the ends A and C and negative charge migrating toward B and D. The charge motion ceases when the downward-pointing internal field E_{int} due to the charge accumulation within the pores exactly cancels the upward-pointing external electric field E_V. However the outer ends of the pores are electrically connected to a large volume of electrically conducting sea water and thus must be at the same voltage when equilibrium is established. This may be thought of as an application of the fact that in equilibrium the interior of all conductors is at a single voltage (see Section 2.2) or that the external ends of the ampullae are connected by a low-resistance conductor in the form of the sea water that short-circuits the ends of the pores. The net result is that if the combined length of the pores is L the total voltage $E_V L$ that is induced across them by the externally generated electric field E_V appears across the voltage detector BC. This transfer, such that the total voltage drop induced across the lengths of the pores appears directly across the detector, may be thought of as a form of voltage amplification. These fishes have been shown experimentally to be capable of detecting vertical electric fields smaller than $1\,\mu V m^{-1}$.

The same detector could be used to measure the horizontal component of the Earth's magnetic field, as we discussed above for the single conductor moving in a magnetic field. The voltage across the detector will be given by

$$V_D = E_{int}L = vB_H L \sin(\theta),\qquad\qquad [14.4]$$

where θ is the angle between the direction of motion $\tilde{v}$ in a horizontal plane and the direction of $\tilde{B}_H$, and L is the combined length of the two pores. Swimming in the north–south direction leads to $V_D = 0$. Swimming east–west at a speed of $v = 1\,m\,s^{-1}$ where the horizontal component of the Earth's field B_H is $30\,\mu T$ results in $V_D = 3\,\mu V$ for $L = 0.1\,m$. This is equivalent to detecting a vertical electric field of $30\,\mu V\,m^{-1}$, well within the measured capability of the fish. In fact, like most biological transducers, the biological voltage detector only responds to changes in V_D which have a characteristic period of less than a few tenths of a second and will not detect static voltages. However, the fish needs only to turn in circles in a horizontal plane or wag its head from side to side while swimming to generate time-varying signals that can reveal the direction and size of B_H. Whether this device is used as a magnetic compass is as yet not fully established, but it is clearly a possibility.

Animals that live in air cannot use this mechanism because air is such a poor conductor. Voltages across the conducting pores would still be generated by motion in a magnetic field, but the voltage between the exposed ends of the pores would no longer be zero because the animal is no longer embedded within a conducting fluid.

The voltage across the detector will then no longer be simply related to the voltage differences induced across the pores and will no longer serve as a reliable detector of the magnetic field.

14.4   The magnetite compass

Many animals are known to synthesize magnetite (Kirschvink and Gould 1981) which is a mixed oxide of iron with formula Fe_3O_4. As we shall see, this is capable of forming very powerful magnets which could be used as a compass needle by an animal. In Section 8.1 we mentioned paramagnetic atoms such as iron that possess a permanent atomic magnetic moment because of uncompensated angular momentum due to the spins and orbital motions of their electrons in partially filled electron shells which, in a semi-classical sense, can be regarded as minute current loops. In the iron group of paramagnets, the partially filled electron shell that gives rise to the paramagnetism is the 3d shell which is the outermost orbit after the 4s electrons form covalent bonds with neighbouring molecules. The various orbital states of the 3d electrons correspond to orbitals pointing in different directions relative to the negative ligands that surround the paramagnetic iron atom in the crystal. The crystal can lower its energy by very small mechanical distortions of its crystal structure, which removes any degeneracy in the energies of the orbital states so that all the 3d electrons of the paramagnetic atoms exist in the single, lowest-energy, orbital state. This leaves only the magnetic moment due to the spin angular momentum free to change its orientation, which leads in turn to **spin-only paramagnetism**.

Within a crystal containing paramagnetic atoms there can exist an exchange interaction which aligns the spins of the electrons within the overlapping orbitals of neighbouring atoms. This is essentially an electrostatic force but it acts indirectly through the quantum mechanical requirements on the symmetry of overlapping electron wave functions. It is a strong force which acts only on the spin angular momentum of the electrons. As is to be expected with overlapping wave functions, the force is very short range and can only act between neighbouring paramagnetic atoms or, as in magnetite, through the intermediary of the overlapping polarized electron orbitals of the oxygen ligand situated between the two paramagnetic atoms. Within the crystal there is competition between the disturbing effects of thermal excitations and the aligning effects of the exchange interaction. As the temperature is lowered there comes a time when the aligning effects of the exchange interactions become dominant and the spins of the paramagnetic atoms become ordered. The onset of order is co-operative and thus sudden as the probability of a given atom having its spin direction ordered depends on the number of neighbouring paramagnetic atoms that are already ordered. For magnetite the ordering temperature is 574 °C, and below that temperature the magnetite is said to be in a ferrimagnetic state. The ordered magnetic structure of a magnetite crystal below this temperature depends upon its size.

For very small crystals (Kirschvink and Gould 1981) with major dimension less than $0.05\,\mu\text{m}$ there are not sufficient paramagnetic atoms to bring about the co-operative ordering and the crystal remains paramagnetic in that it has no net magnetic moment when it is not in an external magnetic field. For larger crystals the magnetic properties depend upon the shape of the crystal. For needle-shaped crystals with length much greater than their width, a crystal with major dimension between $0.05\,\mu\text{m}$ and $1.0\,\mu\text{m}$ is magnetically ordered in a single direction throughout its volume and forms a powerful magnet with a magnetic moment per unit volume M of $4.8 \times 10^5\,\text{A}\,\text{m}^{-1}$ at $300\,°\text{K}$. For a less elongated crystal with length twice its width, this homogeneous magnetized state exists when the major dimension is between $0.05\,\mu\text{m}$ and $0.3\,\mu\text{m}$ Such crystals are called **single-domain ferrimagnetic crystals**. As the crystal grows in size, the energy per unit volume, $BH/2$, stored in the magnetic field surrounding the crystal grows, and at some upper size limit the crystal can lower its total energy by forming two domains which are regions with oppositely directed magnetic moments separated by a domain wall. This drastically reduces the magnetic field around the crystal and thus lowers the energy stored in this field, but destroys the ability of the crystal to act as a compass needle. The transition occurs when the resultant lowering of the energy stored in the magnetic field surrounding the crystal is greater than the energy needed to form the domain wall between the two domains. Within a domain wall between two oppositely directed magnetic domains the spins gradually change in direction from that which exists within one domain to that in the other. Such an arrangement costs exchange energy as the spins are no longer parallel to the direction of neighbouring spins. At even greater sizes the crystal may divide into many domains such that its total magnetic moment is very small.

The interaction energy of a single-domain magnetite particle of magnetic moment m situated in a magnetic field B is given in equation [6.7] as $-mB\cos(\theta)$, where θ is the angle between the directions of the magnetic moment $\tilde{m}$ and the magnetic field $\tilde{B}$. The energy is thus $-mB$ for parallel alignment and $+mB$ for anti-parallel alignment. As discussed in Appendix 2, the ratio of the probability P_p of finding the magnetic moment parallel to the field to the probability P_a of finding it anti-parallel to the field is given by the ratio of the Boltzmann functions for these two orientations, so that

$$\frac{P_p}{P_a} = \frac{\exp\left(\frac{+mB}{kT}\right)}{\exp\left(\frac{-mB}{kT}\right)} = \exp\left(\frac{2mB}{kT}\right). \tag{14.5}$$

A single-domain crystal with dimensions in micrometres of $0.1 \times 0.05 \times 0.05$ has a magnetic moment of $1.2 \times 10^{-16}\,\text{A}\,\text{m}^2$ so that at $25\,°\text{C}$ and in a magnetic field of

$50\,\mu T$, the Boltzmann ratio P_p/P_a is about 18.5. Such a crystal could then act as a very reliable compass needle in pointing along the direction of the prevailing magnetic field of the Earth. The amplifying factor in this detector is the alignment within single-domain particles of very large numbers of atomic magnets by the exchange force, so that the magnetic moment of the crystallite is so large as to give reliable readings in very small magnetic fields, even in the face of thermal fluctuations. As we discussed above, the reliability of a magnetite compass cannot be increased by increasing the size of the magnetite crystal because the crystal will eventually spontaneously divide into two or more oppositely directed domains and the strong magnet is lost. To increase the reliability requires the use of many small single-domain crystals in some kind of array such that the magnetic moments remain parallel. One kind of array that is found within some bacteria (see below) is a linear chain of magnetite needles nose to tail, and oriented parallel (Mann *et al.* 1984). The energy of N such crystals, each with a magnetic moment $\tilde{m}$ magnetically aligned in the direction of a magnetic field B, then becomes $-NmB$. The Boltzmann ratio P_p/P_a for a chain of five aligned single-domain magnetite crystals of the dimensions given above, that is free to rotate as a whole, is then greater than 2 million.

Other types of aligned arrays of magnetite needles are possible. If the needles are incorporated in a suitable liquid crystalline fluid that consists of linear molecules that spontaneously tend to align parallel, it has been predicted theoretically (Brochard and de Gennes 1970) and confirmed experimentally (Rault *et al.* 1970) that the lowest-energy state of the magnetite loaded liquid crystal is when the magnetite needles are aligned parallel to the linear liquid crystal molecules. Twisting the magnetite needles away from this parallel alignment increases the strain in the liquid crystal and thus raises its energy. Within a suitable droplet of such material all the magnetite needles remain parallel to each other and to the ordering direction within the liquid crystal as the needles rotate under the influence of externally applied magnetic fields. Such an correlated array of N needles with an energy of $-NmB$ in a magnetic field B could form an extremely reliable compass (Edmonds 1996). In a biological context, the interior of a lipid bilayer forms an ordered molecular array like a liquid crystal. Magnetite needles incorporated within a local region of a cell membrane could form a powerful magnet that could detect the direction of the Earth's magnetic field by rotation of the cell if it is free or the disruption to the integrity of the membrane if the magnetite needles twist while the cell remains tethered. However, to date no direction-correlated arrays of single-domain magnetite needles other than linear chains have been detected in animals.

Several models of an animal compass based upon single-domain magnetite have been proposed (Kirschvink and Gould 1981; Yorke 1979; Edmonds 1992, 1996) but the only case of orientation by magnetite that has been unambiguously verified is that of magnetotactic bacteria (Blakemore 1975). These live in mud at the bottom of the sea and contain within them a chain of magnetite crystals with parallel magnetic moments. Such is the torque exerted by the Earth's magnetic field on this chain of aligned magnetite magnets that the bacterium is constrained to live its life always oriented such that the magnetic moment of the chain is directed along the magnetic field. To a bacterium in the northern hemisphere moving in the direction of the field means moving deeper into the mud, and moving in the opposite direction means moving toward the surface. Such discrimination is important in that to survive the bacteria must avoid the oxygen-rich atmosphere near the mud surface.

Another powerful magnetic material that is known to be synthesized by living systems is greigite, which has the same formula as magnetite except that the oxygen is substituted by sulphur. This is grown by bacteria near sulphur-rich volcanic vents at the bottom of the sea and forms single-domain ferrimagnetic crystals with properties very like those of magnetite. Magnetite crystals have been found in many animals and it is interesting to note that much of this magnetite is in the form of crystals of the size required to form powerful single-domain magnets. It is then tempting to suppose that the magnetism has some purpose. If forming magnetite crystals were simply a method of storing iron, say, then any size of crystal would suffice.

14.5   The free radical magnetic field detector

When two single-electron wave functions, one on each of two adjacent atoms, overlap they create a bonding orbital which then contains two electrons, one from each atom. The rules of quantum mechanics demand that the spins of the two electrons within a bonding orbital are oppositely directed in what is known as the singlet state and which we will indicate as $\langle S \rangle$. No bonding orbital is possible between atomic electron states with parallel electron spins in what is known as the triplet state, which we represent by $\langle T \rangle$, and the energy in this state drops as the atoms separate. When a molecule is exposed to light of particular wavelengths it can result in the sudden rupture of a covalent bond. The two pieces formed are called free radicals. They each retain one electron that contributed to the original bonding orbital and the anti-parallel orientation of the two spins is retained during the very rapid rupture of the covalent bond. The presence of these exposed valence electron orbits results in the two free radicals being very active chemically, which makes it plausible that small changes in the concentration of free radicals may initiate chemical cascades that result in robust biological responses. On the other hand the strong chemical activity of the radicals can cause damage to living systems and as a result most cells have many mechanisms for scavenging or destroying any radicals produced. The lifetime in aqueous solution of free radicals that do not immediately recombine is typically 1–100 ns and only in very special circumstances as long as 1 µs.

We will now briefly list some of the predicted properties of a spin angular momentum in a magnetic field which are needed in order to understand the free radical field detector. The electron has a spin $S = 1/2$ and possesses a magnetic moment $\tilde{m}$ associated with its angular momentum with an amplitude given by

$$m = -\left(\frac{g}{2}\right)\sqrt{S(S+1)}\left(\frac{qh}{4\pi M}\right), \qquad [14.6]$$

where g is a number close to 2, q is the magnitude of the electron charge, M its mass, and h is Planck's constant, $6.626 \times 10^{-34}\,\mathrm{J\,s^{-1}}$. In a magnetic field $\tilde{B}$ the electron has two quantized energy levels with energies U_+ and U_- such that

$$U_+ = +mB = +\frac{g}{2}\left(\frac{qh}{4\pi M}\right)B,$$

$$[14.7]$$

$$U_- = -mB = -\frac{g}{2}\left(\frac{qh}{4\pi M}\right)B.$$

Quantum mechanics predicts that the direction of the spin angular momentum will precess about the direction of the steady field B with a frequency ν_S such that $h\nu_S = (U_+ - U_-)$. From equation [14.7] we deduce that

$$\nu_S = g\left(\frac{1}{2\pi}\right)\left(\frac{q}{2M}\right)B. \qquad [14.8]$$

Semi-classically, if we were dealing with a circulating electron that has angular momentum because of its orbital circulation, we could apply the Larmor theorem as discussed in Section 10.3 and Chapter 16 and predict that, when the magnetic field is applied, the direction of the angular momentum of the orbiting electron will rotate about the direction of the applied field B at the Larmor frequency ν_L given in equation [10.7] by

$$\nu_L = \left(\frac{1}{2\pi}\right)\left(\frac{q}{2M}\right)B. \qquad [14.9]$$

This semi-classical prediction is the same as the quantum mechanical prediction for the precession rate of an **orbital angular momentum** in a magnetic field. The fact that the precession rate for **spin angular momentum** in equation [14.8] is a factor $g \simeq 2$ greater than the prediction for the precession rate for an orbital angular momentum in equation [14.9] is a quantum electrodynamics result that has no classical equivalent.

Let us return to the two free radicals immediately after their formation in a singlet state $\langle S \rangle$ and discuss how small magnetic fields may be detected by this mechanism (Schulten 1982; Brocklehurst and McLaughlan 1996). The radical pair in a singlet $\langle S \rangle$ state with electron spins anti-parallel can immediately recombine to form the original molecule, whereas a radical pair in a triplet $\langle T \rangle$ state with spins parallel cannot recombine. To have an effect on the average lifetime of a radical pair a magnetic field must alter the transition rate between the original $\langle S \rangle$ state and the $\langle T \rangle$ state. We now distinguish between two types of magnetic fields that can act upon the two spins, one on each radical, that formed the bonding orbital between them. The first is an externally applied field. This is the same for both radicals in the pair and will cause the electron spins within both radicals to precess at the same rate. This cannot change the relative direction of the two spins and so cannot affect the $\langle S \rangle$ to $\langle T \rangle$ conversion rate. The second type of magnetic field is a local field due for instance to the fields generated by the nuclear magnetic moment of the proton which forms the nucleus of neighbouring hydrogen atoms. This will be different for both radicals which will then precess at different rates, so that a radical pair with anti-parallel spins may be converted in time to a radical pair with parallel spins. These local fields will enhance the $\langle S \rangle$ to $\langle T \rangle$ conversion rate and so prolong the life of the radical pair by converting a pair with anti-parallel spins which can immediately recombine to a pair with parallel spins which cannot recombine. If an external steady magnetic field is applied that is small in comparison with the internal fields it will have little effect. If, however, the amplitude of the applied magnetic field is increased until it is bigger than the internal fields then the spins will tend to precess at the same rate, which will decrease the $\langle S \rangle$ to $\langle T \rangle$ conversion rate and hence decrease the average life of the radical pair. We would then expect an increase in the average lifetime of the radical pair when an applied external magnetic field becomes comparable to the local internal magnetic fields.

The mechanism described above could in principle act as a biological detector of the amplitude of the prevalent magnetic field if small differences in the average lifetime of the radicals formed by exposure to light of suitable wavelength could lead to measurable effects in the animal. To act as a detector of the direction of the magnetic field relative to the animal at least one of the radical fragments would need to remain always fixed within the animal. One possibility (Schulten 1982) is that the hyperfine interaction between the electron spin of the fixed fragment and the magnetic moment of a neighbouring proton depends strongly upon direction within the animal. Then external magnetic fields could have different effects depending on their direction relative to the axes, fixed in the animal, that define this hyperfine interaction.

The effect of externally applied fields on the lifetime of radical pairs is easy to demonstrate experimentally (Brocklehurst and McLaughlan 1996) for fields of the order of a few millitesla which are 20 to 100 times larger than the Earth's magnetic field. This is as predicted by the mechanism discussed above because a typical effective magnetic field at an electron spin caused by the hyperfine interaction with a neighbouring hydrogen nucleus is a few millitesla in amplitude. The fact that free radical effects can be readily demonstrated in fields as low as a few millitesla is due to the fact that the mechanism relies upon a bifurcation between different possible outcomes determined by quantum mechanical selection rules based upon spin direction. It does not depend upon a comparison between the energy of these spin states due to their orientation in a magnetic field and typical thermal energies, as is the case with many other possible mechanisms, because the orientation of the spins is only very weakly coupled to the random thermal motion.

To detect fields as small as the Earth's magnetic field by the mechanism described above, which leads to changes in the lifetime of a radical when the external field is comparably to the hyperfine fields, would require free radicals with a structure that results in particularly small hyperfine fields due to neighbouring hydrogen nuclei. Recently another mechanism has been described (Timmel *et al.* 1998; Ritz *et al.* 2000) that could change the lifetime of a radical pair in external magnetic fields much smaller than hyperfine fields. This is essentially quantum mechanical in nature. When a small magnetic field is applied, pairs of electron spin energy levels that were degenerate in the absence of the applied field, split into resolved doublets. The semi-classical equivalent of this splitting is that extra modes of precession of the spins of a radical pair become possible and these extra modes will increase the probability of $\langle S \rangle$ to $\langle T \rangle$ conversion, thus prolonging the lifetime of the pair. Thus the theoretical predictions are that very small applied magnetic fields could increase the $\langle S \rangle$ to $\langle T \rangle$ conversion rate and thus increase the life of a radical pair formed in a singlet state, and that increasing the magnetic field further could lead to a decrease in the $\langle S \rangle$ to $\langle T \rangle$ conversion rate and hence a decrease in the radical pair lifetime as the applied field becomes comparable with the effective hyperfine fields.

There are mechanisms other than recombination that limit the lifetime of a radical pair including longitudinal (spin-lattice) and transverse (spin-spin) relaxation times (see Section 16.4). In order to observe any magnetic field effect upon the radical lifetime an appreciable precession of the electron spin directions must take place before the radicals decay. From equation [14.8] we can deduce that to precess through an angle of $360°$ in a magnetic field of $50\,\mu T$, an electron spin requires a time of about $0.11\,\mu s$. This is towards the upper limit of known radical lifetimes. In addition, once formed, the two halves of a radical pair will tend to drift apart due to diffusion. It is for this reason that research aimed at detecting biological effects at very low magnetic fields is focused on radical pairs in systems where the radicals are prevented from diffusing far apart by being tethered by an organic chain or sequestered in a cavity or micelle.

Despite these difficulties with the free radical model of a compass mechanism, it has been much discussed recently because of the experimental finding that many birds and some amphibians can no longer make use of their magnetic compass if the only light available to them is changed in wavelength (Deutschlander *et al.* 1999). One possible wavelength effect with the free radical model is that to create the radical pair, the frequency of the incident light ν, and hence the energy of the incident photon $h\nu$, needs to be bigger than some threshold value. Thus to operate the mechanism might require light towards the blue end of the visible spectrum.

References

Blakemore, R. P. (1975) Magnetotactic bacteria. *Science* **190**, 377–379.

Brochard, F. and de Gennes, P. G. (1970) Theory of magnetic suspensions in liquid crystals. *Journal de Physique* **31**, 691–708.

Brocklehurst, B. and McLaughlan, K. A. (1996) Free radical mechanism for the effects of environmental electromagnetic fields on biological systems. *International Journal of Radiation Biology* **69**, 3–24.

Deutschlander, M. E., Phillips, J. B. and Borland, S. C. (1999) The case for light-dependent magnetic orientation in animals. *Journal of Experimental Biology* **202**, 891–908.

Edmonds, D. T. (1992) A magnetite null detector as a migrating bird's compass. *Proceedings of the Royal Society* **B249**, 27–31.

Edmonds, D. T. (1996) A sensitive optically detected magnetic compass for animals. *Proceedings of the Royal Society* **B263**, 295–298

Kalmijn, A. J. (1982) Electric and magnetic field detectors in elasmobranch fishes. *Science* **218**, 916–918.

Kirschvink, J. L. and Gould, J. L. (1981) Biogenic magnetite as a basis for magnetic field detection in animals. *Biosystems* **13**, 181–201.

Mann, S., Moench, T. T. and Williams, R. J. P. (1984) A high resolution electron microscope investigation of bacterial magnetite. *Proceedings of the Royal Society* **B221**, 385–393.

Paulin, M. G. (1995) Electroreception and the compass sense of sharks. *Journal of Theoretical Biology* **174**, 325–339.

Rault, J., Cladis, P. E. and Burger, J. P. (1970) Ferronematics. *Physics Letters* **A32**, 199–200.

Ritz, T., Salih, A. and Schulten, K (2000) A model for photoreception-based magnetoreception in birds. *Biophysics Journal* **78**, 707–718.

Schulten, K. (1982) Magnetic field effects in chemistry and biology. *Advances in Solid State Physics* **22**, 61–83.

Timmel, C. R., Till, U., Brocklehurst, B., McLauchlan, K. A. and Hore, P. J. (1998) Effects of weak magnetic fields on free radical recombination reactions. *Molecular Physics* **95**, 71–89.

Weindler, P, Wiltschko, R. and Wiltschko, W. (1996) Magnetic information affects the stellar orientation of young bird migrants. *Nature* **383**, 158–1607.

Wiltschko, R and Wiltschko, W. (1995) *Magnetic Orientation in Animals*, Zoophysiology Volume 33. Springer, Berlin.

Yorke, E. D. (1979) A possible magnetic transducer in birds. *Journal of Theoretical Biology* **77**, 101–105.

15. An electrostatic model of a proton/ion or an ion/ion coport or counterport

Although it is clear that all biological function at the molecular level must be explicable in terms of electric fields and forces, many researchers, who are used to mechanical models, find it difficult to imagine how such an electrical system might operate. In this chapter an electric model of a proton/ion trans-membrane counterport is described as an example of such a system. First the equivalence of the electrical model to the more familiar mechanical model is shown, and then the operation of the counterport explained in some detail. The results of a Monte Carlo computer simulation, based upon the principle of detailed balance discussed in Appendix 2, are given which shows the model to have many of the properties measured with real counterports.

15.1 The ionic coport and the counterport

An ionic coport is a trans-membrane protein or group of proteins that uses the drop across the cell membrane of the electrochemical potential of one type of ion to power the transfer of a second type of ion across that membrane in the same direction but against its trans-membrane electrochemical potential gradient. Any such simultaneous transfer can only occur spontaneously if the Gibbs free energy of the combined process falls, as described in Appendix 3. For definiteness we consider a coport that uses the transfer of the cation X^+ across the membrane, resulting in a change $-\Delta\mu_X$ in its electrochemical potential, to power the transfer of another type of cation Y^+ across the membrane in the same direction when the change in its electrochemical potential is $+\Delta\mu_Y$. If, as a result of this transfer, no other changes have occurred in the Gibbs free energy of this coport, then for a **spontaneous** dual transfer we require that

$$+\Delta\mu_Y - \Delta\mu_X \leq 0. \qquad [15.1]$$

An ionic counterport provides similar coupled transfers across the membrane but the ion transfers now occur in opposite directions.

Such ionic coports and counterports are essential for biological function. If a concentration difference in a single type of ion (say, protons) has been created across the cell membrane by a primary ion pump which uses metabolic energy, then, using coports and counterports, the energy stored in this concentration difference can be used to create many of the other trans-membrane ionic concentration differences required for the cell to function. Without coports and counterports the cell would need to be equipped with an individual primary pump for every type of ion that establishes a concentration difference across the membrane that stores energy.

15.2   A simple mechanical model of a counterport

A simple mechanical model of an ionic counterport (Vidavar 1966; Jardetsky 1966) that is illustrated in many biochemical textbooks is shown in Fig. 15.1. It consist of a mechanical cavity that can open sequentially to the fluid on the outside or on the inside of a cell. Within the cavity are ionic binding sites that can selectively bind type X^+ or type Y^+ cations depending upon the state of the counterport. Let us assume that the transfer of an X^+ type cation across the membrane from outside to inside results in a drop in its electrochemical potential and that the transfer of a Y^+ type ion from inside to outside requires a rise in its electrochemical potential as discussed above. In Fig. 15.1a

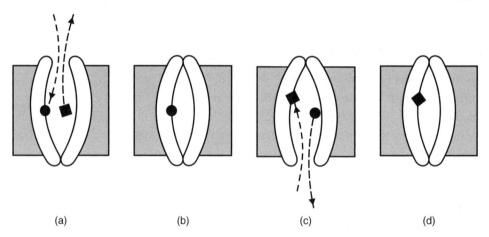

| (a) | (b) | (c) | (d) |

Fig. 15.1   The four states of a popular mechanical model for an ion/ion counterport. (a) In this state a protein spanning the membrane opens a cavity to the upper side and simultaneously creates a binding site for round type ions. (b) A mechanical change causes the cavity to close. (c) Here the protein changes shape again such that the cavity is open to the lower side and the binding site for round type ions is removed. Simultaneously a binding site for square type ions is created. The result is that the previously bound round ion is delivered to the lower side and the cavity binds a square type ion from the lower side. (d) A mechanical change closes the cavity again. (a) The cavity returns to its initial state open to the upper side. The binding site for square type ions is removed and a binding site for a round type ion is created. The complete cycle results in the delivery across the membrane of a round type ion downward and a square type ion upward.

the cavity is open to the outside with an available internal binding site highly selective to (round) ions of type X^+. When the ion binds we assume that the counterport undergoes a shape change such that the cavity is closed, as sketched in Fig. 15.1b. Now the cavity undergoes another change so that it is open to the inside of the cell. Simultaneously the binding site selective to X^+ type ions ceases to exist but another site highly selective to (square) ions of type Y^+ is created. The X^+ type ion is thus expelled from the binding site and finds its way into the internal fluid, while a Y^+ type ion from the inside binds to its selective site within the cavity, as illustrated in Fig. 15.1c. To complete the cycle, when the Y^+ ion binds, the counterport closes its cavity (Fig. 15.1d) and then subsequently returns to the state with the cavity open to the outside fluid. Simultaneously with this structural change, the internal binding site for type Y^+ ions ceases to exist and an internal site highly selective to ions (round) of type X^+ is created. The (square) Y^+ type ion is discharged to the outer fluid and eventually an X^+ type ion (round) binds to the internal site as shown in Fig 15.1a, to complete the cycle.

   The net result of the cycle is the transfer of a single Y^+ type ion from the inside to the outside using the drop in the electrochemical potential of the X^+ type ion transferring from the outside to the inside. As a physical model of a counterport this mechanical model leaves many questions unanswered. For example, how is the change in the internal binding site coupled to the mechanical shape of the cavity, and how is the Gibbs free energy drop resulting from the transfer of the X^+ ion stored such that it can be utilized to power the sequential transfer of the Y^+ type ion against its electrochemical gradient? The strength of the model is as a teaching aid which clearly separates the change in access from the cavity to both sides of the membrane, and the change in site selectivity necessary to distinguish between the two types of ion.

15.3   An electrostatic analogue of the mechanical model for a proton/ion counterport

We consider (Edmonds and Berry 1991) an ion channel highly selective for a particular type of cation X^+. Buried in the protein surrounding the aqueous pore of the channel are two acid sites S_1 and S_2 close to the pore and equally distributed across the membrane. We assume further that a proton path exists linking these sites with each other and with the bathing solution at both sides of the membrane, as indicated by the zigzag line in Fig. 15.2. We will denote the charged states of the acid sites as $(0, 0)$ if both sites are uncharged, $(1, 0)$ and $(0, 1)$ if the left or right site is charged, and $(1, 1)$ if both sites are charged. The pK_a of the acid sites buried in an environment that is only slightly electrically polarizable within the protein are chosen to be greater that the pH of the solution bathing the membrane so that (see Section 11.3 and 11.5) the acid sites normally exist in their uncharged state.

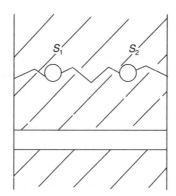

Fig. 15.2 An ion channel penetrates across a membrane. In the proximity of the channel are two acid sites S_1 and S_2 which are connected together and to the bath at each side of the membrane by a proton conducting path indicated by the zigzag line.

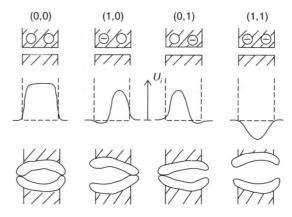

Fig. 15.3 The four states of an electrical model of the proton/ion counterport shown in Fig. 15.2. The four states differ in the charged states of the two acid sites. For example, the state $(1,0)$ indicates that the left-hand acid state is in a negative state, having lost a proton. Below the depiction of each state is shown a graph of the energy of a univalent cation passing through the ion channel plotted against its position in the channel. At the bottom of the diagram is shown for each state the equivalent state of the mechanical model depicted in Fig. 15.1.

In Fig. 15.3 we sketch the four possible states of the acid sites and below each sketch is a graph of the energy of a X^+ cation in the pore plotted against its position within the pore. Below these graphs are sketched the equivalent state of the mechanical model discussed above. In the $(0, 0)$ state of the counterport, the electrostatic self-energy of the X^+ ion U_i is high anywhere within the narrow pore (see Section 13.2) and the probability of an ion entering the pore from either side is small. The small graph below the sketch of the $(0, 0)$ configuration shows the high self-energy of the cation X^+ plotted against its position within the pore. The $(0, 0)$ state is analogous to the mechanical counterport when the cavity is closed at both ends, as sketched below this graph in the figure. If the left-hand acid site should lose a proton to the left-hand solution, it becomes negatively charged and the state changes to $(1, 0)$. The electric field of the negatively charged acid site penetrates the pore and lowers the energy of a cation in the left-hand end of the pore as sketched in Fig. 15.3 so that the pore resembles the mechanical cavity open on the left but remaining closed on the right. A X^+ type cation is then free to enter the pore from the left. If it does so, it stabilizes the acid site S_1 in its negatively charged state because of the electric field of the cation acting on the left-hand acid site (see Section 11.4). The negative charge on the acid stabilizes the presence of the cation in the left-hand end of the pore and the presence of this cation in turn stabilizes the charged state of the acid $(1, 0)$ so that this combination of charged acid site and cation in the left-hand end of the pore is locked together until the cation moves.

An exactly similar situation exists if the right-hand acid site loses a proton to the right-hand solution to be come negatively charged so that the state $(0, 1)$ prevails. This state corresponds to the mechanical model open to the right-hand solution but remaining closed to the left-hand solution. If both acid sites become charged, the electric field in the pore lowers the energy U_i of cations in the pore, as shown by the right-hand graph for the state $(1, 1)$ in Fig. 15.3. This state resembles the mechanical cavity simultaneously open at both ends. Such a state could allow the uncoupled leakage of ions across the membrane. However, the repulsive electrostatic interaction between the two acid sites when both are negatively charged raises the electric energy of the whole counterport in the state $(1, 1)$. As we shall see below in the computer simulation, we arrange that the electrostatic repulsion between the two acid sites is such that the state $(1, 1)$ becomes highly unlikely and contributes little to the operation of the counterport.

Thus it can be seen that the states $(0, 0)$, $(1, 0)$ and $(0, 1)$ are analogues of the states of the mechanical model in which the cavity is closed at both sides of the membrane, open on the left but closed on the right and open on the right but closed on the left. These states are brought about by electrostatic forces and require no mechanical motion other that the transfer of protons to and from the acid sites. Because only proton motion is required, the changes can be very fast. The dual selectivity in this model counterport is arranged by having the first ion (the proton) travelling along a different path than the X^+ cation.

15.4 Kinetics of the model

We consider now how the state of the model may change, starting from the state $(1, 0)$ with a X^+ cation occupying the left-hand end of the pore. The first possibility is that the cation returns to the left-hand solution, from which it came, thereby removing the stabilization of the charged state of the site S_1. If the S_1 site loses its charge by binding a proton from the left-hand solution, the model has returned to its initial state $(0,0)$ with no cations in the pore. If, however, before the S_1 site loses its negative charge, another X^+ cation enters the pore from the left, the model returns to the state $(1,0)$ with a X^+ cation in the left-hand end of the pore as at the start of this paragraph. A second possibility is that the X^+ cation moves towards the right within the pore. This lowers the probability that S_1 remains charged and increases the probability that S_2 becomes charged. If the movement to the right of the cation X^+ continues, the state may change from $(1,0)$ to $(0,1)$ by the transfer of a proton from the site S_2 to the site S_1. This change in the charged states of the acid sites makes the configuration with the cation X^+ at the right-hand end of the pore particularly stable. The cation can now leave the channel on the right, removing the stabilization of site S_2 in its charged state, and the site S_2 may return to its uncharged state by binding a proton from the solution on the right. Should the sequence described immediately above occur, it results in the coupled transfer of a X^+ cation from left to right and the net transfer of a proton from right to left as expected for a proton/ion counterport.

However, after every step in the sequence described above for the counterport action, there are other possibilities for changes of state of the acid sites and motions of the X^+ ion within the channel, so that the number of possibilities becomes very large. Figure 15.4 describes the allowed transitions between the states of the acid sites

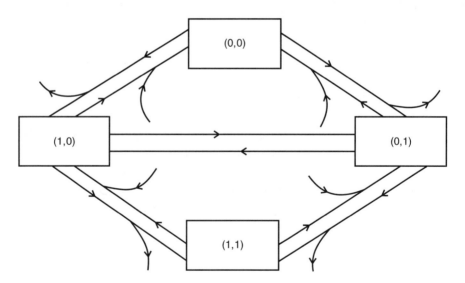

Fig. 15.4   A diagram showing the possible transitions between the four states of the electrical model of the proton/ion counterport shown in Fig. 15.3. The curved arrows indicate proton exchange with the aqueous solution bathing the faces of the membrane.

(0, 0), (1, 0), (0, 1) and (1, 1). The curved arrows in this kinetic diagram represent the binding of a proton by an acid site from the solution bathing the membrane or the transferring of a proton from an acid site to the solution. Thus, in the transition between the state (0, 1) and the state (0, 0), the curved arrow represents the binding of a proton from the right-hand solution to the acid site S_2, whilst in the transition between the states (0, 0) and (0, 1) the curved arrow represents the transfer of a proton from the site S_2 to the right-hand solution. The straight arrows between the states (1, 0) and (0,1) represent the transfer of a proton directly between the two acid sites. To make the situation even more complex, the probabilities of transfer between any two states in Fig. 15.4 depend upon the position of the cation X^+ within the pore. To model a counterport it is necessary to make the state (1, 1) unlikely because of strong electrostatic coupling between the two acid sites when both are charged as described above. Thus the most probable transitions are confined to the upper half of Fig. 15.4. Even with this simplification and a second of only allowing a single X^+ cation in the pore at any particular time, this time due to cation–cation repulsion within the pore, the situation remains very complicated with a vast number of possible trajectories, and the only satisfactory way of discovering how the state of the whole counterport would evolve with time is by computer simulation as described below.

15.5   A Monte Carlo computer simulation

The full description of the simulation, with references to previous work, is given in the original paper (Edmonds and Berry 1991) but here we will describe the essential features and present some results of the simulation. To simplify the simulation, the path of the ion X^+ through the pore is represented by five steps. Step 1 is just outside the pore on the left, step 5 is just outside the pore on the right and steps 2, 3 and 4 are evenly distributed along the length of the pore. The acid sites S_1 and S_2 are situated within the width of the membrane such that they experience three-quarters and one-quarter of the total membrane voltage $V(\text{left}) - V(\text{right})$ that is applied across the membrane. The pK_a values for the two acid sites were taken as 8.5, and the electrostatic interaction between the charged acid sites and the cation at step N is given in multiples of a typical thermal energy kT as $U(S_1, S_2, N)$, where (S_1, S_2) gives the state, such as (0, 1), of the acid sites and

$$U(1, 0, 2) = U(0, 1, 4) = -18kT, U(0, 1, 2) = U(1, 0, 4) = -5kT,$$

$$U(1, 0, 3) = U(0, 1, 3) = -18kT,$$

$$U(S_1, S_2, 1) = U(S_1, S_2, 5) = 0 \quad \text{for all} \quad (S_1, S_2). \tag{15.2}$$

The self-energy of the X^+ cation within the pore was taken as $+18kT$, and the interaction between the two acid sites when both are charged was taken as $+15kT$. Finally, the trans-membrane voltage is fixed together with the concentration of the cations and the pH of the fluids bathing each side of the membrane. The cation concentration and the membrane voltage contribute to the predicted probability that a cation will enter an empty pore from either side, and the two values of pH together with the membrane voltage contribute to the probability that a proton will transfer between the acid sites and the bathing solutions.

The proton kinetics are assumed to be much faster than those of the cation. At each tick of an imaginary clock the position of any cation in the pore is recorded. When this is known the energies of the various charged states of the acid sites are known and thus the transition probabilities in Fig. 15.4 are known using detailed balance (see Appendix 2). The transition probabilities depend upon the membrane voltage assumed, and those which involve the transition of a proton between the sites and the bathing solutions (curved arrows in Fig. 15.4) also depend upon the pH assumed for the two bathing solutions. The most probable state of charge of the acid sites can then be determined using the method of partial diagrams (Hill 1966). Having determined the charged state of the acid sites, the energy of the cation at any step is known. The free energy of the whole system with the cation in a given position, is then calculated in this initial state as $U_{initial}$. This free energy must include (1) cation–acid site interactions, (2) acid site–acid site interaction, (3) the self-energy of the cation within the pore and (4) the interaction of both the acid sites and the cation with the assumed trans-membrane voltage. The total free energy of the system U_{final} may then be calculated with the cation in each of the possible neighbouring positions along its path. The probability of a transition between the initial state and each of the possible final states may now be written as $P(i \rightarrow f)$ in the expressions below using the **principle of detailed balance** described in Appendix 2:

$$P(i \rightarrow f) = \begin{cases} \dfrac{1}{2}\exp\left[\dfrac{-(U_{final} - U_{initial})}{kT}\right] & \text{if } U_{final} > U_{initial} \\[2mm] \dfrac{1}{2} & \text{if } U_{final} \leq U_{initial} \end{cases} \qquad [15.3]$$

The equations express the fact that if $U_{final} = U_{initial}$ for ionic movement to the left or the right there is an equal probability of 1/2 for movement in either direction. However, if the movement in a given direction involves the surmounting of an energy barrier ΔU, then that probability must be 1/2 multiplied by the Boltzmann probability $\exp(-\Delta U/kT)$ that at temperature T the barrier will be surmounted. Thus the probabilities P_L, P_R, P_S of moving to the left, of moving to the right or staying in the same position (not surmounting the barrier) may be found such that $P_L + P_S + P_R = 1$.

To determine if and how the cation moves under the influence of thermal fluctuation at temperature T, a Monte Carlo technique is employed. Imagine a straight line that stretches from 0 to 1 and that the line is divided into three sections. The first section stretches from 0 to P_L, the second has length P_S and stretches from P_L to $P_L + P_S$, and the third stretches from $P_L + P_S$ to 1 and thus has a length P_R. If a computer-generated random number RAN such that $0 \leq RAN \leq 1$ is created it may be laid along the line from 0 to RAN. If RAN falls within the first section of the line with length P_L a move to the left is indicated, if RAN falls within the third section of length P_R a move to the right is indicated, and if RAN falls within the second section of length P_S then no motion is indicated. It can then be seen that if this is repeated many times the calculated motion becomes stochastic or random but the statistics of the moves determined in this way and averaged over time are consistent with the probabilities determined by detailed balance. For a particular simulation, the ionic concentrations and the pH values of the solutions bathing the two sides of the membrane are fixed, as is the trans-membrane voltage. At each tick of an imaginary clock the charged state of the acid sites is first determined as described above, the probabilities for ion motion are then calculated and a random number is generated which determines whether the ion moves and if so in which direction along its path. The system is then allowed to evolve freely over millions of timing clock ticks and a count is kept of the number of cations and protons that cross the membrane in both directions.

Figure 15.5a shows the predicted functioning of the model counterport when a favourable proton gradient, which causes proton flow from left to right, drives a cation flux from right to left in many cases against the prevailing ionic electrochemical potential gradient. The graph shows the ion flux through the pore from right to left when the concentration of cations is equal on both sides of the membrane but where there exists a proton gradient from left to right of 10:1 ($\Delta pH = pH_L - pH_R = 1$), shown as open squares, and 100:1 ($\Delta pH = 2$), shown as open circles. The cation flux is measured as the net number of cation transfers that occur from right to left in 1 million clock ticks and is plotted against the membrane voltage $V_m = V(\text{left}) - V(\text{right})$ assumed for each simulation. At all positive membrane voltages when $V(\text{left}) > V(\text{right})$ all cation transfer from right to left requires a free energy input and could not occur spontaneously. Figure 15.5b shows the stoichiometry ratio of proton transfers to cation transfers, which is very close to unity and almost independent of the assumed membrane voltage. This shows that only about 5% of the proton transfers occur uncoupled to a simultaneous cation transfer. Thermodynamically it is known that the stoichiometry ratio may not be exactly unity, so that some ions leak through the pore, but the model shows very tight coupling between the oppositely directed proton and ion flows.

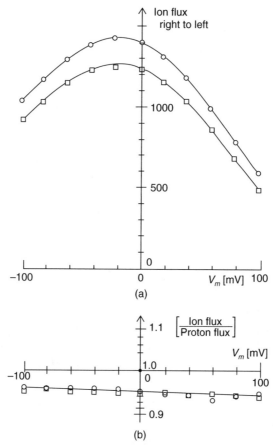

Fig. 15.5   (a) A graph of the predicted net cation transfer passing from right to left for 1 million cycles of the electrical counterport model, plotted against the trans-membrane voltage. The cation concentration is the same in the solution bathing each side of the membrane. These transfers are driven by a proton concentration difference from left to right of 10:1 (open squares) or 100:1 (open circles) in the solutions bathing the membrane faces. (b) A graph of the ratio of ion transfer number to the proton transfer number plotted against the trans-membrane voltage difference. This demonstrates the tight coupling of the oppositely directed flows which is almost independent of the trans-membrane voltage difference.

Figure 15.6a shows that the model counterport is reversible and that a cation gradient can drive protons across the membrane against the prevailing proton electro-chemical potential gradient. The full lines in Fig. 15.6a show the predicted right to left proton flux when the pH is 7 in both bathing solutions but when there is a cation concentration ratio of 100:1 (open circles) and 10:1 (open squares) from left to right. Again the transfer is measured as the net proton transfer from right to left in 1 million clock ticks. Figure 15.6b shows the stoichiometric ratio which reveals that, regardless of membrane voltage, less that 5% of ion transfers can occur without a simultaneous proton transfer in the opposite direction. The broken line in Fig. 15.6a shows the

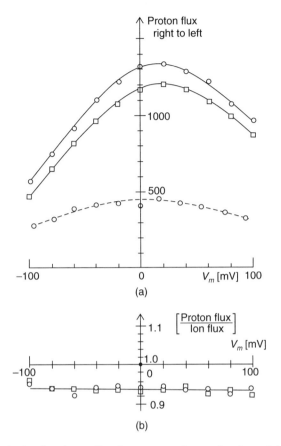

Fig. 15.6 (a) A graph showing the predicted proton transfer passing from right to left for 1 million cycles of the electrical counterport model, plotted against the trans-membrane voltage. The proton concentration is the same with pH = 7 for the solution bathing each side of the membrane. The proton flow is driven by a cation concentration ratio left to right of 10:1 (open squares and full line) or 100:1 (open circles and full line) in the solutions bathing the membrane faces. The broken line gives the predicted proton transfer when the left to right ion concentration is 100:1 but when the driven protons face an unfavourable proton ratio right to left of 1:10. (b) A graph of the ratio of proton transfer number to ion transfer number plotted against the trans-membrane voltage. The coupling ratio is essentially independent of the membrane voltage.

driven right to left proton flux when there is also an adverse 10:1 proton concentration ratio so that $pH_L = 6.5$ and $pH_R = 7.5$. Thus even when both the membrane voltage and the proton concentration ratio are unfavourable for proton transfer, the favourable cation ratio can drive the protons across the membrane.

We have in this chapter shown that a very simple electrical model exhibits behaviour similar to that found in nature and that it provides an analogue to popular mechanical models illustrated in textbooks. It requires no mechanical motion of the trans-membrane proteins that form the counterport and it will operate very fast as it requires only ionic migration. With a different set of parameters chosen a proton/ion coport may be modelled. A slight elaboration of the model in which two neighbouring ion channels, which are selective for different types of ion, are situated close to the same chain of ionizable acid or basic residues can be shown to give rise to correlated ionic transfers (an ion/ion coport or counterport) in the two channels (Berry and Edmonds 1992, 1993). As in the model above, the ions passing through the parallel channels interact with each other because of the effects of their electric fields on the charged states of the ionizable residues situated between the two channels. No direct ion-to-ion electrical coupling or mechanical motion of the protein is required. Such models can show the kinetic characteristics normally associated with carriers (Berry and Edmonds 1992) or with channels that can contain several ions simultaneously (Berry and Edmonds 1993).

References

Berry, R. M. and Edmonds, D. T. (1992) Carrier-like behaviour from a static but electrically responsive model pore. *Journal of Theoretical Biology* **154**, 249–260.

Berry, R. M. and Edmonds, D. T. (1993) Correlated flux through parallel pores: Application to channel subconductance states. *Journal of Membrane Biology* **133**, 77–84.

Edmonds, D. T. and Berry, R. M. (1991) The proton ladder, a static mechanism for ion/proton coports and counterports. *European Biophysics Journal* **20**, 241–245.

Hill, T. L. (1966) Studies in irreversible thermodynamics (IV). Diagrammatic representation of steady state fluxes for unimolecular systems. *Journal of Theoretical Biology* **10**, 442–459.

Jardetsky, O. (1966) Simple allosteric model for membrane pumps. *Nature* **211**, 969–970.

Vidavar, G. A. (1966) Inhibition of parallel flux and the augmentation of counterflow by transport models not involving a mobile carrier. *Journal of Theoretical Biology* **10**, 301–306.

16. An introduction to the semi-classical theory of pulsed nuclear magnetic resonance

As an illustration of the Larmor theorem described in Chapter 10, the semi-classical theory of nuclear magnetic resonance (NMR) is introduced in this chapter. The essential resonant mechanism is explained such that an oscillating but small-amplitude magnetic field can rotate a nuclear magnetic dipole moment even when it is acted upon by a much larger-amplitude static magnetic field. The concepts of the rotating frame, the 90° pulse, the 180° pulse, the relaxation times T_1 and T_2 and the spin echo are described in some detail. The uses of NMR as a detector of microscopic molecular structure and also of the macroscopic internal structure of humans for medicine are briefly introduced.

16.1 Classical angular momentum and the Larmor theorem

Nuclear magnetic resonance is probably the most used analytical technique in chemistry and biology. It is concerned with the manipulation, by the application of magnetic fields, of the magnetic dipole moment associated with the nucleus of a particular type of atom within the molecule studied. The nuclear magnetic dipole moment may be thought of as due to a microscopic electric current loop created by a charged nucleus that has angular momentum and therefore spins. There is no exact classical analogue of the angular momentum associated with the quantum mechanical concept of spin, so that we will first predict the behaviour of the angular momentum and magnetic dipole moment associated with a single charged particle circulating in a loop, and later point out any differences between this classical orbital angular momentum and that created by a quantum mechanical spin.

We first consider classically the motion of a particle of mass m and charge q rotating in a small plane circle of radius R such that the normal to the plane of the circle makes an angle α to the z-axis, as sketched in Fig. 16.1a. At time t, let the radius vector from the origin to the particle make an angle $\theta(t)$ with a fixed diameter of the circle and let the frequency with which the particle rotates be given by f so that the angular frequency of rotation in radians per second is given by $\omega = d\theta/dt = 2\pi f$. The velocity

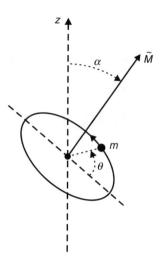

Fig. 16.1   A particle of mass m and charge q circulates in a plane circle such that the normal to the circle makes an angle α to the z-axis. The resultant magnetic moment created by the moving particle is shown as $\tilde{M}$.

of the particle moving around the circle in polar coordinates is $v = Rd\theta/dt$, as described in Appendix 1, so that its momentum directed round the circle is mv. Its angular momentum $\tilde{G}$ is directed along the normal to the circle and has an amplitude of

$$G = Rmv = mR^2\frac{d\theta}{dt} = mR^2 \times 2\pi f. \qquad [16.1]$$

The magnetic dipole moment of the rotating particle $\tilde{M}$ is given by equation [6.3] as the product of the area of the circle (πR^2) and the equivalent current I that circulates around the circle. Now a charge q passes a given point on the circle f times each second so that we have $I = qf$. Thus the magnetic dipole moment of the rotating particle is given by

$$M = \pi R^2 qf. \qquad [16.2]$$

An important parameter for magnetic resonance is the ratio of the magnetic dipole moment M and the angular momentum G. This ratio M/G is called the **magnetogyric ratio** and is usually represented by the symbol γ, so that, using equations [16.1] and [16.2], we obtain

$$\gamma = \frac{M}{G} = \frac{q}{2m}. \qquad [16.3]$$

The Larmor theorem, discussed in Section 10.3, predicts that if a fixed magnetic field of amplitude B_0 is applied along the z-axis, the motion of the particle will remain the same except that the circular orbit and its normal will precess about the z-axis with the Larmor angular velocity ω_L such that

$$\omega_L = -\frac{q}{2m}B_0 = -\gamma B_0. \qquad [16.4]$$

When we are concerned with a magnetic moment $\tilde{M}$ associated with the **spin angular momentum** of a nucleus $\tilde{G}$, the Larmor theorem still predicts that the spin will precess about the magnetic field B_0 with an angular frequency $\omega_L = -\gamma B_0$. However, the magnetogyric ratio γ for a spin differs from that of a classical orbital angular momentum in that, for the spin, $\gamma = g_N q/2m$, where g_N is a number called the **nuclear g-value** which has a value close to 2. A true quantum mechanical treatment would also take account of the fact that the angular momentum is quantized so that its projection on the z-axis may only take on particular values. We will continue with the simpler classical treatment which provides a more immediate insight into the operation of pulsed nuclear magnetic resonance.

16.2   The rotating frame

We now introduce the concept of a **rotating frame**. The 'laboratory frame' $Oxyz$ is, as its name suggests, fixed in the laboratory and its z-axis and origin O are shown in Fig. 16.2. We imagine another set of Cartesian axes $Ox'y'z'$ with the same origin as the laboratory axes and with the Oz' axis coincident with the $O'z$ axis but where the Ox' and the Oy' axes rotate about the Oz axis in the same direction as the Larmor precession but with an angular velocity ω, as shown in Fig. 16.2. The axes $Ox'y'z'$ form the rotating frame. Let us imagine that we observe the motion of the nuclear magnetic dipole

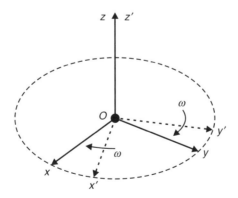

Fig. 16.2   A set of Cartesian axes shown as $Oxyz$ is fixed in space and is called the laboratory frame. A second set of Cartesian axes $Ox'y'z'$ shares the same z-axis but rotates about its z'-axis with an angular velocity ω; this second set of axes is called the rotating frame.

moment while we, as observers, are fixed in the rotating frame. If $\omega = 0$ then the rotating frame is the same as the laboratory frame and we observe the motion in the magnetic field unchanged, with the magnetic moment of the nucleus precessing about the z-axis with angular frequency $\omega_P = \omega_L = -\gamma B_0$. If, however, the angular frequency with which the rotating frame rotates is equal to the Larmor angular frequency so that $\omega = \omega_L$, then the magnetic moment of the nucleus appears fixed and does not precess relative to the rotating frame, so that the precession frequency relative to the rotating frame becomes $\omega_P = 0$. If the rotating frame rotates with angular frequency ω in the same direction as ω_L and has a magnitude $|\omega|$ intermediate between $|\omega_L|$ and 0 such that $0 \leq |\omega| \leq |\omega_L|$, then the magnetic moment precession rate relative to the rotating frame becomes $\omega_P = \omega_L - \omega$. To an observer fixed in this rotating frame, rotating at angular frequency ω, the magnetic moment appears to precesses at the angular frequency $\omega_P = \omega_L - \omega$ predicted by the Larmor theorem for a magnetic field of $B_0 + \omega/\gamma$ because

$$\omega_P = \omega_L - \omega = -\gamma \left(B_0 + \frac{\omega}{\gamma} \right). \tag{16.5}$$

The quantity $B_0 + \omega/\gamma$ is called the **effective magnetic field in the rotating frame** and we will denote it by B_E. That we can define a meaningful effective magnetic field in this manner is a direct consequence of the Larmor theorem which equates the presence of a magnetic field with a simple rotation of the whole picture about the direction of the magnetic field, with no other significant changes to the motion.

It is important to be clear about the algebraic signs of the parameters we have defined. For a positively charged nucleus γ is positive, as shown in equation [16.3]. But if the magnetic field B_0 is applied along the positive z-axis, the Larmor precession corresponds to a negative rotation about the positive z-axis, as discussed in Section 10.3, so that ω_L is negative, as can also be seen from equation [16.4]. The angular frequency of the rotating frame ω is chosen in the same direction as the Larmor frequency and thus is also negative. Therefore ω/γ in equation [16.5] is negative so that the effective field in the rotating frame $B_E = B_0 + \omega/\gamma$ is smaller that B_0. This is of course consistent with the fact that the spin precession angular frequency ω_P appears to be lower if we observe it from a rotating frame that rotates in the same sense as the precessing spin.

As this is the essence of the understanding of the classical theory of magnetic resonance, we will restate how the precession of the nuclear magnetic moment in a magnetic field B_0 applied along the positive z-axis appears to an observer fixed in the rotating frame. When the angular rotation frequency of the rotating frame ω is zero then $B_E = B_0$ and the motion observed by an observer fixed in the rotating frame, is the same as that seen from the laboratory frame, with an effective field of B_0 acting along the z-axis and an effective precession rate $\omega_P = \omega_L = -\gamma B_0$. When $\omega = \omega_L$, no precession is observed by an observer fixed in the rotating frame because then $B_E = 0$ so that the magnetic moment observed from the rotating frame is static and the nucleus appears to behave as it would in the absence of an applied magnetic field. When ω is in the same direction as ω_L and has a magnitude $|\omega|$ less than $|\omega_L|$ such that $0 \leq |\omega| \leq |\omega_L|$, then in our coordinate systems both ω_L and ω are negative so that $B_E = B_0 + \omega/\gamma$ is smaller than B_0 and the observed precession rate of the nuclear magnetic moment is given by $\omega_P = \omega_L - \omega$ and has a magnitude $|\omega_P|$ less than $|\omega_L|$.

16.3   Application of a small-amplitude rotating magnetic field and magnetic resonance

We now consider the effect of applying a small amplitude steady magnetic field $B_1 \ll B_0$ along the direction of the Ox' axis of the rotating frame so that it rotates in the xy plane of the laboratory frame with a negative angular frequency ω about the Oz axis. If $\omega = \omega_L$ then, observed from the rotating frame, there is no magnetic field along the Oz', axis, but, fixed in this frame, there is a magnetic field B_1 along the Ox' axis. The Larmor theorem will then predict that the magnetic moment of the nucleus will precess about the Ox' axis at an angular frequency $\omega_1 = -\gamma B_1$. Seen from the laboratory frame, the motion is complex and is the superposition of a fast precession about the Oz axis at an angular frequency $\omega_L = -\gamma B_0$ and a much slower precession about the Ox' axis of the rotating frame at an angular frequency $\omega_1 = -\gamma B_1$.

Before we applied the rotating magnetic field B_1, the nuclear magnetic dipole moment $\tilde{M}$ made an angle α with the Oz axis of the laboratory frame as in Fig. 16.1 and it thus had a magnetic energy, according to equation [6.7], of

$$U_M = -MB_0 \cos(\alpha). \tag{16.6}$$

However, because of its precession about the rotating field B_1, the angle that the magnetic moment $\tilde{M}$ makes with the magnetic field $\tilde{B}_0$ changes with time and, after a rotation of π about the Ox' axis, it will now make an angle of α with the negative direction of the z-axis, equivalent to an angle $\pi - \alpha$ with the magnetic field B_0. At this time its magnetic energy has increased to $U_M = +MB_0 \cos(\alpha)$.

This effect illustrates a major feature of the magnetic resonance phenomenon. By applying a magnetic field of small amplitude B_1 that rotates at the Larmor frequency, we can change the direction of a magnetic dipole moment from nearly parallel to a large magnetic field B_0 to nearly anti-parallel to the large field and thus substantially change its magnetic energy even though $B_1 \ll B_0$. It is this ability of a very small magnetic field to twist a magnetic dipole moment which is acted on by a very much bigger magnetic field that is at the heart of magnetic resonance. The amplitude of the small rotating magnetic field B_1 can in fact be arbitrarily small in comparison with the large static magnetic field B_0 and the effect will still occur. However, the twisting is governed by the Larmor precession about the Ox' axis with angular frequency $\omega_1 = -\gamma B_1$ so that the rate of twisting is proportional to the amplitude of the magnetic field B_1. One reason why the small rotating magnetic field B_1 is so effective in twisting the magnetic dipole moment at resonance when $\omega = \omega_L$, is that it is fixed in the rotating frame which is rotating at the Larmor frequency and thus remains fixed relative to the precessing magnetic moment.

If we apply the rotating magnetic field B_1 along the Ox' axis of a rotating frame which is rotating at an angular frequency less than the Larmor frequency so that $|\omega| < |\omega_L|$, then, to an observer in this frame, there is an effective static magnetic field $B_0 + \omega/\gamma$ pointing along the Oz axis and a second static magnetic field of amplitude B_1 pointing along the Ox' axis. Thus, to an observer in this frame, the total effective magnetic field has an amplitude of

$$B_T = \sqrt{\left[\left(B_0 + \frac{\omega}{\gamma}\right)^2 + B_1^2\right]},$$

it lies in the $Ox'z'$ plane of the rotating frame and it makes an angle β with the Oz axis such that

$$\tan(\beta) = \frac{B_1}{\left(B_0 + \dfrac{\omega}{\gamma}\right)}.$$

The Larmor theorem then predicts that the magnetic dipole moment will precess about this total field B_T with an angular frequency $-\gamma B_T$. Because $B_0 \gg B_1$ the angle β will be very small unless ω is very close in magnitude to ω_L and is in the same rotational sense, so that $B_E = B_0 + \omega/\gamma$ is close to zero. Thus this theory of magnetic resonance, based solely on classical mechanics, predicts that the rotating small-amplitude magnetic field will have little effect upon the precession of the magnetic dipole moment around the direction of the large static field B_0 unless $B_0 + \omega/\gamma$ is comparable in magnitude to B_1, or equivalently unless ω is close to ω_L. When ω is close to ω_L the magnetic dipole moment will precess about the total field B_T in the rotating frame and when $\omega = \omega_L$ the magnetic dipole moment will precess about the direction of the rotating field B_1 (the Ox' axis) at an angular rate $-\gamma B_1$ so that it continuously oscillates from making an angle α to making an angle $\pi - \alpha$ with the large field B_0 along the z-axis.

16.4   The detection of nuclear magnetic resonance, the 90° pulse and the free precession signal

The simplest equipment for the detection of magnetic resonance consists of three coils with their axes pointing along the three perpendicular Cartesian axes. A large and powerful solenoid, usually wound with superconducting wire, provides a static field of fixed amplitude B_0 directed along the z-axis of the laboratory frame. A small transmitter coil has its axis along the x-axis of the laboratory frame and is tuned to the Larmor frequency $f_L = \omega_L/2\pi$ of the nucleus to be studied in the steady field B_0. A third detector coil is tuned to the same Larmor frequency as the transmitter coil and has its axis parallel to the y-axis of the laboratory frame. The transmitter and detector coils are located at right angles to minimize the direct magnetic field coupling between the transmitter and detector coils. The transmitter coil produces a sinusoidal, radio-frequency, linearly polarized magnetic field $2B_1 \cos(\omega_L t)$ along the x-axis of the laboratory frame.

The linearly polarized magnetic field along the x-axis which varies sinusoidally in length at an angular frequency ω_L is equivalent to two magnetic fields, each of fixed amplitude B_1, which rotate in opposite directions about the z-axis with angular frequencies $+\omega_L$ and $-\omega_L$. This can be seen in Fig. 16.3. At time $t = 0$ we assume that the two rotating fields point along the x-axis to the right and thus add vectorially to produce a field of amplitude $2B_1$ along the x-axis. At the time t, illustrated in the figure, the two fields of amplitude B_1 point to the right and their vector sum is a linear field of length $2B_1 \cos(\omega_L t)$ pointing along the x-axis to the right. At a later time, when the two rotating fields point directly upward and downward, the resultant field is zero in amplitude. Still later, when the rotating fields point to the left, their vector sum is a linear field pointing to the left along the negative x-axis. In the figure the vertical

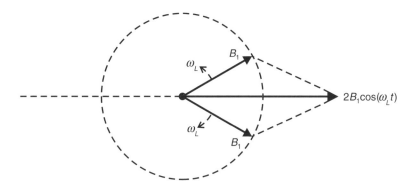

Fig. 16.3   This diagram shows the equivalence of two magnetic fields of fixed magnitude B_1 which rotate in opposite directions with the same angular velocity ω_L and a single magnetic field of magnitude $2B_1 \cos(\omega_L t)$ which oscillates in amplitude along a single direction.

components of the rotating fields always cancel to zero and their vector sum is thus always horizontal along the x-axis. The two rotating fields of fixed length B_1 add vectorially to produce a linear field in a fixed direction but with a length $2B_1 \cos(\omega_L t)$. These arguments show the equivalence of the two rotating fields of fixed strength and a linear field which points along a fixed direction but varies in strength sinusoidally.

The magnetic field component of fixed amplitude B_1 which rotates at the Larmor frequency rotates in synchronism with the nuclear magnetic moments that have this Larmor frequency and is responsible for the magnetic resonance as described above. In the rotating frame it is a small magnetic field of fixed amplitude B_1, directed along the Ox' axis. The field component which rotates at $-\omega_L$ is rotating at a frequency different by $2\omega_L$ from the rotating nuclear magnetic moment and is so far from synchronism or the resonance condition as to have a negligible effect upon the motion of the nuclear moment. Thus the static transmitter coil with its axis along the Ox axis of the laboratory frame provides the magnetic field B_1 directed along the Ox' axis of the rotating frame described in the theory of magnetic resonance discussed above.

We will now consider the effect of applying short pulses of radio frequency excitation to an assembly of identical nuclei in an applied steady magnetic field of B_0 each with a magnetogyric ratio γ and thus a Larmor frequency of $\omega_L = -\gamma B_0$. The radio frequency excitation at the Larmor frequency ω_L is applied to the transmitter coil for a short period of $(\pi/2)(1/\gamma B_1)$ seconds. In the frame $Ox'y'z'$, rotating with B_1 at an angular frequency ω_L, there is no effective field along the Oz' axis but a small-amplitude fixed field of amplitude B_1 is directed along the Ox' axis during the pulse. The precession frequency of the nuclear magnetic moments about the Ox' axis is $-\gamma B_1$ and during the pulse they will rotate through an angle of exactly $\pi/2$ or $90°$. If at the start of the pulse the nuclear magnetic moments were aligned along the Oz axis, with a minimum energy in the field B_0, then at the end of the pulse they will be aligned along the Oy' axis of the rotating frame. In the laboratory frame the nuclear magnetic moments which were originally aligned along the z-axis will rotate together into the Oxy plane and there precess about the z-axis with an angular frequency ω_L. These rotating magnetic moments will induce in the detector coil a radio-frequency emf and hence a radio-frequency current at frequency ω_L according to Faraday's law, expressed by equation 9.1. A radio-frequency excitation at the Larmor frequency that is transmitted for a time τ_{90} given by

$$\tau_{90} = \frac{\pi}{2}\left(\frac{1}{\gamma B_1}\right) \qquad [16.7]$$

is called a **90° pulse** because, when a rotating field B_1 is applied along the Ox' axis for this time, it rotates those nuclear magnetic moments with magnetogyric ratios of γ about the Ox' axis through a precession angle of $90°$.

This is the essence of nuclear magnetic resonance. The signal in the detector coil immediately after the end of the 90° pulse is proportional to the number of nuclei in the sample that have a magnetogyric ratio of γ. Nuclei with different magnetogyric ratios will not respond appreciably, so that we have an analytical technique for measuring the number of nuclei of a given type within the irradiated sample. The signal in the detector coil will rapidly decay in amplitude for two distinct physical reasons. The first is that there will be microscopic differences in the applied magnetic field at different locations in the sample, due both to imperfections in the magnet which generates the steady magnetic field B_0 and to slight variations in the field at a given site because of the magnetic field generated by the nuclear magnetic moments of neighbouring nuclei. As a result the nuclear magnetic moments that all pointed along the Oy' axis immediately after the termination of the 90° pulse as sketched in Fig. 16.4a will precess at slightly different rates about the Oz axis. As a result they will 'fan out' in the (x, y) plane as they precess as sketched in Fig. 16.4b until they are equally distributed in the (x, y) plane around the Oz axis when the magnetic field they generate at the detector coil is constant in time and the emf generated falls to zero as predicted by Faraday's law. In Fig. 16.4b the fastest precessing moment is marked F and the slowest S. This decay in the detected signal is characterized by a **transverse relaxation time** T_2 such that the detected signal amplitude $S(t)$ at time t is given by

$$S(t) = S(0) \exp\left(\frac{-t}{T_2}\right), \qquad\qquad [16.8]$$

where $S(0)$ represents the maximum signal immediately after the 90° pulse.

The second physical mechanism involves the changes in energy of the nuclear magnetic moments in the large steady field B_0. We assume that before the 90° pulse is applied, each of the nuclear moments $\tilde{M}$ are aligned approximately along the z-axis of the laboratory frame and have an energy close to the minimum energy of $-MB_0$. We assume that the nuclear moments within the specimen are in thermal equilibrium at the temperature T, the absolute temperature of the specimen. If α is the angle between the direction of the nuclear magnetic moments and the z-axis then, after the 90° pulse, $\alpha = 90°$ and their magnetic energy $-MB_0 \cos(\alpha)$ is raised from $-MB_0$ to 0 in the field B_0. They are thus now much hotter than the rest of the specimen and will tend to cool through thermal contact with the rest of the specimen. Cooling involves the reduction in the angle α from 90° to an angle close to zero such as they had before the 90° pulse was applied. This gradual motion of the rapidly precessing nuclear magnetic moments from along the Oy' axis of the rotating frame after the pulse to along the Oz axis again reduces the signal induced in the detector coil, and this signal loss is characterized by a **longitudinal relaxation time** T_1 such that

$$S(t) = S(0) \exp\left(\frac{-t}{T_1}\right). \qquad [16.9]$$

Of course in reality both relaxation processes occur simultaneously such that the nuclear moments, which are together all aligned along the Oy' axis of the rotating frame after the 90° pulse, simultaneously fan out and move towards the Oz axis so that at an intermediate state they are fanned out on the surface of a circular cone symmetrically disposed about the z-axis. The electrical signal registered in the detector coil immediately after the 90° pulse is called **the free precession signal**.

16.5   The 180° pulse and the spin echo

The invention of the **spin echo** by Prof. Erwin L. Hahn of the University of California represented a major advance both in the practical application and in the theory of magnetic resonance. It relies upon the application of both a 90° pulse as described above and a 180° pulse with twice the duration of the 90° pulse such that all the nuclear magnetic moments in the specimen with a particular magnetogyric ratio precess about the Ox' axis of the rotating frame by 180°. First a 90° pulse is applied at the Larmor frequency of the chosen set of nuclei to the transmitter coil and then, at a time τ_S after the start of the 90° pulse, a 180° pulse is applied at the same frequency. With no further pulses applied, the spin echo signal appears spontaneously at a time τ_S after the start of the 180° pulse. For simplicity of explanation we will consider the case when $T_2 \ll T_1$ so that the free precession signal is reduced to zero predominantly due to the fanning out of the nuclear moments precessing at slightly different frequencies in the (x, y) plane, although the technique is applicable in all cases. As seen in Fig. 16.4a, immediately after the 90° pulse all the selected nuclear spins with a given γ point along the Oy' axis in the rotating frame. A short time after the application of the 90° pulse, the spins have begun to fan out due to the slight differences in the applied magnetic field B_0 at different locations. Figure 16.4b shows the spin marked F which is precessing the fastest and that marked S which is precessing the slowest, with all the other spins precessing at intermediate rates. Seen by an observer in the laboratory frame, the spins that were initially pointing along the Oz-axis have now formed a fan in the (x, y) plane which rotates about the z-axis at angular frequency ω_L and becomes wider as time passes. At a time τ_S after the 90° pulse a 180° pulse is applied which rotates the fan of spins around the Ox' axis of the rotating frame by 180°. Now the slowest spin is precessing in advance of the fastest spin, as shown in Fig. 16.4c. If the spins maintain their precession rates the fan will grow narrower with time until at

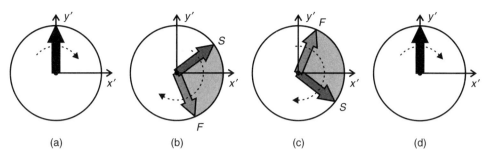

Fig. 16.4   An illustration of the spin echo. A group of spins are located in a large but static magnetic field directed along the z-axis which varies very slightly from spin to spin. The initial state of the magnetic moments of the spins is aligned along the static field direction. (a) By applying a 90° pulse along the Ox' axis of the rotating frame at the Larmor frequency for the spins, they rotate to point along the Oy' axis of the rotating frame. (b) Because the spins experience slightly different static fields along the z'-axis they precess around that axis at slightly different rates. The fastest precessing spin is shown as F and the slowest as S and the other spins form a continuous fan between these positions. (c) A 180° pulse is now applied along the Ox' axis which rotates the fan of spins by 180° about that axis. The slowest precessing spin S is now precessing in advance of the fastest precessing spin F. (d) At a time after the 180° pulse equal to the separation in time between the 90° and the 180° pulses, the spins are bunched together as in the state shown in (a) and are thus capable of exciting the maximum signal in the detector coil.

a time τ_S after the 180° pulse all the spins will again precess together as they did immediately after the initial 90° pulse, as sketched in Fig. 16.4d. This is the time of the maximum detected signal, after which the fan will again widen until the signal is lost as it was after the initial 90° pulse. The usual analogy is to a group of runners who run around a circular track each maintaining a constant speed. At some time after the common start a gun is fired, at which point the runners reverse direction but continue to run at the same rate until they all cross the original starting line together.

One matter of practical importance of the technique is that the spin echo signal is detected at a time of our choosing after all the applied radio-frequency pulses are over and the effects on the detection equipment of these pulses has died out. A second practical advantage is that the effects of slight inhomogeneities in the magnetic field B_0 due to magnet imperfections no longer reduce the peak signal amplitude. On the theoretical side the spin echo makes us think carefully about the mechanisms involved in magnetic resonance. For example, the appearance of the ordered bunch of spins that comes together to create the spin echo signal after the same spins have apparently become disordered and the signal lost may appear to violate the entropy principle in which all spontaneous processes must move in a direction in which the disorder increases or remains the same, as discussed in Appendix 2. That the principle is not violated is explained by the fact that the spins remain highly ordered throughout the process, with each precessing at a determined fixed rate depending upon the local value of B_0. The detected signal is lost as the spins fan out but the order remains so that the signal may be revived. Note that this means that the fanning out of the spins due to local magnetic fields caused by the magnetic moments of neighbouring spins is not recovered by the spin echo. The neighbouring magnetic moments fluctuate because they are subject to thermal energy. Thus the excess magnetic field at one spin caused by the magnetic moments of neighbouring spins fluctuates randomly during the spin echo. The fanning out of the spins due to this random process is genuine disorder that is not corrected by the spin echo as predicted by the entropy principle. In a liquid specimen this contribution to T_2 is much reduced because of the rapid tumbling of the molecules which tends to average to zero the total fluctuating internal magnetic field created by the near neighbours of a particular molecule.

16.6   Nuclear magnetic resonance as a structural technique on a molecular scale

Nuclear magnetic resonance, and particularly the resonance of the hydrogen nuclei within an organic molecule, can yield detailed information about the molecular structure of that molecule (Gadian 1995). An important advantage of NMR over X-ray diffraction as a structural technique in biology is that a single crystal of the molecule is not required. The structure of, say, a protein that forms part of an trans-membrane ion channel may be studied while it occupies its biologically functional position buried within the membrane. What is required to observe NMR is an array of identical molecules with immediate environments that are identical. The molecules, however, need not form a regular array or be aligned as in a crystal. A disadvantage of nuclear resonance is that the intrinsic sensitivity is quite low because of the small size of the nuclear magnetic moment, so that very many examples of the molecule studied must be included in the specimen. However, technical advances have been rapid and the determination of the full three-dimensional structure of molecules the size of a small protein is now possible using only NMR.

Let us imagine that we are studying the NMR of the hydrogen nucleus or proton within an organic molecule. Rather than detecting a single resonance it is found that the resonances are spread over a narrow band of frequencies and the shape and distribution of these frequencies can yield detailed information about the structure of the molecule studied. The chief mechanism responsible for the spread in the observed frequencies is called the **chemical shift**, which has its origin in the inter-action of the magnetic moments of the protons and the electron orbits that compose the molecule. In zero field with a diamagnetic molecule there is no extra field at the protons due to the electron orbits. However, in the large applied magnetic field B_0 the electron orbits may be distorted so as to change the magnetic field experienced by a proton within the molecule from B_0 to $B_0 + \Delta B_0$, where ΔB_0 is a small (a few parts per million) correction to the magnetic field experienced by the proton. One example of the chemical shift is the magnetic shielding of a proton in the vicinity of a benzene ring. The alternating single and double carbon to carbon bonds around the ring means that the ring is capable of conducting electron currents with no resistance around the ring. When such a ring is placed in a magnetic field an emf is induced around the ring, as predicted in equation [9.1], such that the direction of the current induced produces a steady magnetic field which subtracts from the original applied magnetic field. Thus protons near such a ring experience a 'shielded' magnetic field which is slightly smaller than the applied magnetic field. This effect may also be considered as an example of a type of diamagnetism as discussed in Section 10.4. As a result of such chemical shifts in the apparent Larmor frequency of the protons in different positions within a molecule, a catalogue which correlates chemical shifts with given molecular environments may be compiled. Thus a given distribution of frequency-shifted proton magnetic resonance spectral lines may act as the structural fingerprint of a particular molecular structure.

Another important interaction that yields structural information is the spin-spin interaction between neighbouring nuclei. This can be of two forms, direct or 'through the air' and indirect or 'through the bond'. The direct coupling is just the magnetic field of one magnetic dipole (see Fig. 6.1a and equation [6.4]) interacting with the magnetic dipole moment of the other (see the problems for Chapter 6). The indirect interaction is caused by the slight distortion of an electron orbit interacting with one nuclear moment and the effect of this on a neighbouring nuclear moment. As an example of spin-spin interaction when studying the nuclear resonance of protons, two neighbouring protons with the same chemical shift can give rise to a characteristic triplet spectrum with the central line of twice the intensity of the outer lines. Other structural techniques include the substitution of deuterium for hydrogen at particular locations within the molecule followed by a comparison of the proton resonance spectra measured for the normal and the substituted molecule. Yet another technique consists of measuring the complete proton spectrum of a given molecule while simultaneously irradiating strongly at the known Larmor frequency of one particular chemically shifted proton. The induced rapid flipping of the irradiated proton can cancel to zero the effects of spin-spin coupling between this proton and its neighbours. The interpretation of complicated NMR spectra has been systematized and automated by computer programs so that a whole variety of effects may be incorporated into the interpretation of the spectrum. As a result increasingly complicated molecules have had their structures determined by NMR.

16.7 Nuclear magnetic resonance as a structural technique on a macroscopic scale

An increasingly frequent use of NMR is the magnetic resonance imaging (MRI) of parts of the human body in medicine to obtain a non-invasive picture of its internal structure (Pykett 1982). With very few exceptions it is the proton resonance that is employed because of the large nuclear magnetic moment of the proton and the common presence of hydrogen atoms in the organic molecules and of water within the body. The emphasis nowadays is on minimizing the time that the patient needs to remain immobile within the magnet, and to this end many ingenious and complicated pulse sequences are employed, with many extra magnetic field gradient pulses applied between the 90° and 180° of the spin echo sequence. We will here be concerned only with establishing the principles of the technique even if the methods we describe would involve unacceptably long expose times for the patient.

We imagine a set of Cartesian coordinates with the z-axis along the axis of the solenoid which provides the uniform static field B_0. We also apply an additional field in the z-direction ΔB_Z which varies with z such that the total magnetic field along the z-axis is given by

$$B_Z = B_0 + \Delta B_Z = B_0 + Az, \qquad [16.10]$$

where A is a constant. A transmitter coil then creates a rotating field of strength B_1 which rotates around the z-axis with the proton Larmor angular frequency ω_L as usual. If we choose

$$\omega_L = -\gamma(B_0 + Az_1), \qquad [16.11]$$

where γ is the magnetogyric ratio for protons, then only the protons within a narrow sheet of the irradiated specimen perpendicular to the z-axis and with a z-coordinate close to z_1 will contribute to the spin echo signal. The amplitude of the signal will then reflect the number of protons within this narrow sheet. The experiment is then repeated with a different value of ω_L so that the number of protons in a narrow sheet at a different value of z_1 is measured. In this manner the number of protons in equal-width narrow sheets may be measured and stored in a computer such that such sheets when stacked together cover the whole volume of the irradiated specimen.

In an exactly similar manner the experiment is repeated with the excess z-directed steady field varying with the x-coordinate so that $\Delta B_Z = Cx_1$, where C is a constant, so that the total magnetic field along the z-axis is given by

$$B_Z = B_0 + \Delta B_Z = B_0 + Cx_1 \qquad [16.12]$$

and $\omega_L = -\gamma(B_0 + Cx_1)$. In this manner the number of protons in a thin sheet perpendicular to the x-axis such that x close to x_1 is measured and stored. By repeating the experiment with different values of x_1 the proton density in such sheets that cover the whole irradiated body may be obtained. Finally, the numbers of protons within equal-width narrow sheets perpendicular to the y-axis of the irradiated specimen are similarly measured and stored. Using this stored information, the number of protons within a small volume at any position (x, y, z) within the irradiated specimen is known, which provides a three-dimensional map of the specimen in terms of proton density. Without further experiment the density of protons at any place within the patient may be recalled. The proton density within narrow sheets in any orientation can then be readily constructed and examined. Note that it is not possible, in order to save time, to apply static field gradients along, say, two axes simultaneously because this is only equivalent to applying a single static gradient along a third direction.

An alternative strategy that is less dependent on the precise stability of both B_0 and the applied angular frequency ω_L is to apply the fixed static field B_0 along the z-axis together with a sinusoidally time-varying field gradient ΔB_Z in the same direction such that the total magnetic field along the z-axis becomes

$$B_Z = B_0 + \Delta B_Z(t) = B_0 + A(z - z_1)\cos(\omega_G t). \qquad [16.13]$$

If the transmitter angular frequency is $\omega_L = -\gamma B_0$, where ω_L is the Larmor frequency of the proton in a steady field B_0, then only a narrow sheet of the specimen perpendicular to the z-axis near $z = z_1$ will contribute to the spin echo signal. Elsewhere the Larmor frequency of the protons will differ from ω_L because at any time t the magnetic field $B_0 + \Delta B_Z(t)$ deviates from B_0 so that these protons will not be in resonance, and, in addition, the extra magnetic field $\Delta B_Z(t)$ varying with angular frequency ω_G will tend to broaden in frequency, and hence smear out, any weak proton resonances that do occur. By applying magnetic fields along the z-axis such as

$$B_Z = B_0 + A(x - x_1)\cos(\omega_G t),$$

$$B_Z = B_0 + A(y - y_1)\cos(\omega_G t),$$

the proton density in narrow sheets perpendicular to the other two Cartesian axes may be measured. In this manner the three-dimensional map of proton density within the whole specimen may be computed. Note that unlike the first technique, two or three oscillating field gradients with amplitudes that vary along two or three of the Cartesian directions may be applied simultaneously to determine in a single experiment the number of protons within a linear volume or around a single point.

A particularly interesting recent development of MRI has been the detection of which parts of a subject's brain are most active in real time while the subject carries out some known task (Raichle 1994). An important feature of these experiments is that the maps of increased neuronal activity are superimposed upon a simultaneously obtained, accurate MRI proton density map of the structure of the particular brain being studied. Such experiments are leading to a rapidly increasing understanding of the larger-scale architecture of the brain. The technique relies upon the fact that the haemoglobin in the blood supply that is not carrying oxygen is paramagnetic (see Section 8.1) and thus carries a large electronic magnetic dipole moment, but oxygen carrying oxyhaemoglobin is diamagnetic and carries a much smaller electronic magnetic moment. The fluctuating local magnetic field caused by the rapid, thermally activated, flipping of the paramagnetic dipole moment of the haemoglobin shortens the relaxation time of the protons in this region. This can be detected in the spin echo signal so that differences in the decay rate of the proton signal measured can be used to measure the ratio of haemoglobin to oxyhaemoglobin in a given region. When the neurons in a given area of the brain are active, the fresh oxygenated blood supply to that part increases. Thus an increase in the ratio of oxygenated to deoxygenated haemoglobin in a given region measured by MRI, signals that the region has increased its neuronal activity. It is not yet known why the oxygenated blood supply increases immediately there is increased neuronal activity because, strangely enough, active neurons do not require more oxygen when active for short periods as they then act effectively anaerobically with a local store of oxygen. The fact that neurons behave in this manner was one of the first discoveries made when applying MRI techniques to the brain.

References

Gadian, D. G. (1995) *NMR and Its Application to Living Systems*. 2nd edition, Oxford University Press, Oxford.
Pykett, I. L. (1982) NMR imaging in medicine. *Scientific American* May, 54–64.
Raichle, M. E. (1994) Visualizing the mind. *Scientific American* April, 36–42.

Appendix 1: Mathematics

A1.1.   Cartesian and polar coordinates

Most students are familiar with **Cartesian** coordinates as sketched in two dimensions in Fig. A1.1. Any position $P(x, y)$ in this plane is uniquely defined by specifying two numbers (x, y) which measure displacements along the two perpendicular x- and y-axes. The displacement increments are dx directed along the x-axis and dy directed along the y-axis. The general rule for generating the elementary increments in any coordinate system is to hold all the coordinates constant except one, which is allowed to increase slightly. In Cartesian coordinates the displacement increments are independent in that dx may be defined without knowing y and vice versa. Because the

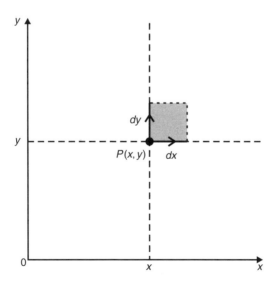

Fig. A1.1   Two-dimensional Cartesian coordinates. A position in the plane $P(x, y)$ is specified by two numbers (x, y). Along the horizontal axis or abscissa the distance is specified by x and along the vertical axis or ordinate the distance is specified by y. The infinitesimal increments in the coordinates are dx and dy, so that the increment in area is $dxdy$.

Cartesian axes are all at right angles to each other, the increment in area is given by *dxdy* in two dimensions and the increment in volume is given by *dxdydz* in three dimensions. All differential increments are assumed to be very small. In fact, strictly speaking, any mathematical expression involving these increments is only valid as we take the limit when the size of all the increments tends towards zero. The rates of change with time of the space coordinates can represent velocities. These are obtained by dividing the space increments by the increment in time *dt* and taking the limit as *dt* tends to zero so that differential quantities are obtained. The velocities directed along the *x*- and *y*-axes are thus *dx/dt* and *dy/dt* respectively. Cartesian coordinates are particularly easy to use and to visualize. Sadly, rectangular shapes are very seldom found in nature, and other coordinate systems that better fit the more common circular, spherical and axially symmetric shapes found are more often used.

Polar coordinates are commonly used for describing situations that have spherical or cylindrical symmetry. Figure A1.2 shows the field point $P(R, \theta)$ in two dimensional spherical polar coordinates. R is the distance from the origin of coordinates at O to the point P, and θ is the angle between the z-axis and the line from the origin that passes through the point P. Following the rule given above for the generation of the distance increments, we first keep θ constant and allow R to increase by a small amount dR to create one increment of length dR which points along the direction of R. Then we keep

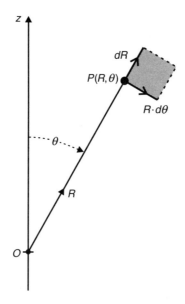

Fig. A1.2   Two-dimensional polar coordinates. A position in the plane $P(R, \theta)$ is specified by giving two numbers (R, θ). The distance from the origin is R and the angle between the z-axis and the direction of $\tilde{R}$ is θ. The infinitesimal increments are dR and $Rd\theta$ so that the increment in area is $RdRd\theta$.

R constant and increase θ to $\theta + d\theta$ to create the other increment. This causes the field point $P(R, \theta)$ to move along the circumference of a circle of radius R and centred at the origin. By definition, the angle $d\theta$, measured in radians, subtended at the centre of a circle of radius R by an arc of the circumference of length L is given by $d\theta = L/R$. The distance moved by the point F when creating the second increment is therefore $L = Rd\theta$. Thus, in two-dimensional polar coordinates, the distance increments are dR and $Rd\theta$. The directions of these increments are commonly referred to as the 'R' and the 'θ' directions. Because $d\theta$ is very small we incur a negligible error if we assume that $Rd\theta$ is a straight line rather than a curve and that the increment $Rd\theta$ is perpendicular to the increment dR. Thus the increment of area dA becomes $dA = dR(Rd\theta) = RdRd\theta$. Note that in these coordinates the two distance increments are not independent in that in the θ-direction, the increment $Rd\theta$ depends upon both R and θ.

Dividing by dt and taking the limit as dt tends to zero generates the velocity components v_R and v_θ in the R and θ directions respectively given by

$$v_R = \frac{dR}{dt} \quad \text{and} \quad v_\theta = R\frac{d\theta}{dt}.$$

In situations that have axial symmetry the full three-dimensional picture may be described by two-dimensional polar coordinates like those shown above. The full three-dimensional picture is then obtained by rotating the whole graph about the z-axis.

A1.2.   The work done by forces and couples

The **work done by a force** $\tilde{F}$ when it causes motion is defined as the magnitude of the force multiplied by the component of the displacement of the point of application of the force in the direction of the force. Figure A1.3 shows a force $\tilde{F}$ acting at a point P

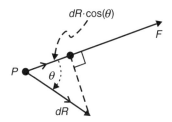

Fig. A1.3   The force acting at the position P is $\tilde{F}$ and the displacement of the point P is $d\tilde{R}$ which makes an angle θ with the direction of $\tilde{F}$. By definition the work done dW by the force $\tilde{F}$ during the displacement $d\tilde{R}$ is given by the amplitude of the force F multiplied by the projection of $d\tilde{R}$ on to the direction of $\tilde{F}$ which is $dR \cdot \cos(\theta)$ so that $dW = FdR \cdot \cos(\theta)$.

and causing a small displacement dR which need not be in the direction of the force $\tilde{F}$. The direction of the displacement dR makes an angle θ with the direction of the force. The component of the displacement in the direction of the force is therefore given by $dR \cdot \cos(\theta)$. The work done dW by the force $\tilde{F}$ during this displacement dR is defined to be

$$dW = FdR \cdot \cos(\theta).$$

If the displacement is in the direction of the force ($\theta = 0$) then $\cos(\theta) = 1$ and the work performed by the force is simply given by $dW = FdR$. To find the total work W done by the force during a macroscopic displacement we must sum the increments in work dW over all the incremental displacements dR that define the total motion of the point of application of the force between the initial and final points in space, represented by a and b. This summation is written as the line integral (see Section A.6)

$$W = \int_a^b dW = \int_a^b F \cos(\theta) dR.$$

A **couple** consists of two forces of equal magnitude F which act in anti-parallel directions but are not collinear. Because the forces are equal in magnitude but opposite in direction they cannot lead to the translation of any body they act upon, but they do exert a twisting action. The magnitude of the couple C, which measures the twist exerted, is defined as $C = Fd$ where d is the distance between the directions of the two forces measured perpendicular to these directions as in Fig. A1.4a. If, under the influence of the forces, the direction of d rotates by a small angle $d\theta$, then the point

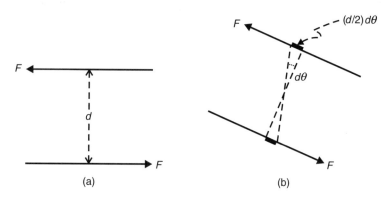

Fig. A1.4   (a) A couple consists of two equal forces of equal amplitude F acting in antiparallel directions when a distance d apart. A couple exerts no net translational force but exerts a twisting motion measured by the moment of the couple $C = Fd$. (b) When a couple rotates by a small angle $d\theta$ the work done dW by its two forces is given by $dW = 2F(d/2)d\theta = Fd\, d\theta = Cd\theta$.

of application of each force F moves through a small arc of length $(d/2)d\theta$ as shown in Fig. A1.4b. Once again, because $d\theta$ is so small, we make a negligible error by assuming that the arcs of length $(d/2)d\theta$ are straight lines and by assuming that the cosine of the very small angle between these arcs and the directions of the forces F is equal to 1. Thus the work dW done by the two forces that comprise the couple in this rotation is given by $2F(d/2)d\theta$. We therefore deduce that the work done by a couple C when it rotates through a small angle $d\theta$ is given by $dW = Fdd\theta = Cd\theta$.

As with the force, the total work done by a couple W when it twists through the finite angle between $\theta = \theta_1$ and $\theta = \theta_2$ is calculated using a line integral so that

$$W = \int_{\theta_1}^{\theta_2} dW = \int_{\theta_1}^{\theta_2} Cd\theta.$$

A1.3. Vectors

A **vector** represents some quantity such as an electric field which has at any point in space an **amplitude** and a **direction**. In this book a vector is represented by a symbol such as $\tilde{v}$ with a wavy line or tilde above it. The amplitude or length of this vector may be printed as $|\tilde{v}|$, which is called the **modulus** of the vector $\tilde{v}$, or it may be printed as v, the symbol for the vector without the line above it. The direction of the vector may be printed as $\hat{v}$, which is the symbol for the vector surmounted by a shallow inverted v or caret. $\hat{v}$ has the direction of the vector but an amplitude or length of 1. Thus we can represent the vector by separating its amplitude and direction:

$$\tilde{v} = |\tilde{v}| \cdot \hat{v} = v \cdot \hat{v}.$$

Consider a vector which starts at the origin of a set of three-dimensional Cartesian axes. The vector can be defined in length and direction by giving the x-, y- and z-**components of the vector**. The components are scalar numbers which represent the projection of the vector on the three Cartesian axes. Thus for a vector $\tilde{v}$ with components (v_x, v_y, v_z) we may write $v_x = v \cdot \cos(\theta_x)$, where θ_x is the angle between the vector and the x-axis, with similar expressions for v_y and v_z. The vector is often written in terms of its components as

$$\tilde{v} = (v_x, v_y, v_z).$$

Applying Pythagoras's theorem successively to the components it is easily shown that the square of the amplitude of the vector $\tilde{v}$ is given by

$$v^2 = v_x^2 + v_y^2 + v_z^2.$$

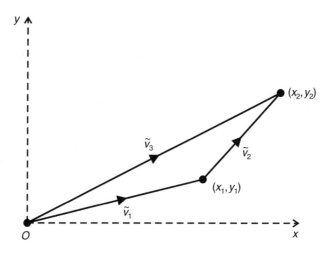

Fig. A1.5 The addition of two vectors $\tilde{v}_1$ and $\tilde{v}_2$ to yield an equivalent resultant vector $\tilde{v}_3$. The vectors to be added are drawn nose-to-tail in sequence and the resultant vector is that which goes from the start of the first vector to the end of the last vector.

If two vectors $\tilde{v}_1$ and $\tilde{v}_2$ act at the same point we can find a single vector $\tilde{v}_3$ which is equivalent to the combined effect of the two original vectors. Let $\tilde{v}_1$ start at the origin 0 of a set of three-dimensional Cartesian coordinates and end at the point (x_1, y_1, z_1). We place the second vector $\tilde{v}_2$ so that it starts at (x_1, y_1, z_1) and say that it ends at the point (x_2, y_2, z_2). The resultant vector $\tilde{v}_3$ is then the vector that starts at the origin and extends to the point (x_2, y_2, z_2). This is illustrated in two dimensions in Fig. A1.5. If the components of $\tilde{v}_1$ are (v_{1x}, v_{1y}) and the components of $\tilde{v}_2$ are (v_{2x}, v_{2y}) then the x- and y-components of $\tilde{v}_3$ are $(v_{1x} + v_{2x}, v_{1y} + v_{2y})$.

The length and direction of the resultant total vector $\tilde{v}_{tot}$ which is equivalent to the sum of any number of vectors in three dimensions all acting at the same point (x, y, z) can be determined in the same manner by placing the component vectors nose-to-tail in a chain. If the ith vector $\tilde{v}_i$ has components $(v_{x,i}, v_{y,i}, v_{z,i})$, then the x- and y- and z-components of $\tilde{v}_{tot}$, which we write as $(v_{x,tot}, v_{y,tot}, v_{z,tot})$, are given by

$$v_{x,tot} = \sum_i v_{x,i}, \quad v_{y,tot} = \sum_i v_{y,i}, \quad v_{z,tot} = \sum_i v_{z,i},$$

where the summations are taken over all the i vectors. Finally, the resultant vector may be determined in both amplitude and direction as its components along the three chosen perpendicular Cartesian directions x, y and z are now known:

$$\tilde{v}_{tot} = \left(v_{x,tot}, v_{y,tot}, v_{z,tot} \right).$$

A1.4.   Vector products

There are two particularly useful vector products. The **scalar product** or **dot product** between two vectors $\tilde{A}$ and $\tilde{B}$ is written as the two vectors with a dot between them as $\tilde{A} \cdot \tilde{B}$ just as often with scalar multiplication. As its name suggests, the scalar product is a scalar number, N say, such that

$$N = \tilde{A} \cdot \tilde{B} = AB\cos(\theta),$$

where θ is the angle between the positive directions of $\tilde{A}$ and $\tilde{B}$, and A and B are the amplitudes of the two vectors. It represents the projection of the length of one vector on to the direction of the other. The scalar product may be written in terms of the components of the two vectors such that

$$N = \tilde{A} \cdot \tilde{B} = A_x B_x + A_y B_y + A_z B_z.$$

A possible use of a scalar product is the representation of the work done by a force, discussed above and illustrated in Fig. A1.3. If the point of application of a force, represented by a vector $\tilde{F}$, moves through a small distance, represented by a vector $d\tilde{R}$, then the increment of work done is a scalar quantity represented dW in the expression

$$dW = FdR \cdot \cos(\theta) = \tilde{F}d\tilde{R}.$$

The **vector product** or cross product of two vectors $\tilde{A}$ and $\tilde{B}$ is written as $\tilde{A} \times \tilde{B}$ and is itself a vector $\tilde{C}$. The amplitude of $\tilde{C}$ is given by

$$C = |\tilde{A} \times \tilde{B}| = AB\sin(\theta).$$

The angle θ is that between the directions of $\tilde{A}$ and $\tilde{B}$, and A and B are the amplitudes of the two vectors. The direction of $\tilde{C}$ is perpendicular to both $\tilde{A}$ and $\tilde{B}$ and it points in the direction of advance of a right-hand thread rotated from the direction of $\tilde{A}$ to that of $\tilde{B}$. A vector product may be written in terms of the components of the two vectors so that

$$\tilde{A} \times \tilde{B} = \tilde{C} = (C_x, C_y, C_z) = (A_y B_z - A_z B_y, A_z B_x - A_x B_z, A_x B_y - A_y B_x).$$

An example of the use of a vector product is the expression for the couple $\tilde{C}$ which acts on an electric dipole $\tilde{p}$ when it is placed in an electric field $\tilde{E}$ such that the angle between $\tilde{p}$ and $\tilde{E}$ is θ as in equation [4.4]. Thus the amplitude of C is given by

$$C = pE\sin(\theta) = |\tilde{p} \times \tilde{E}|.$$

A1.5. Vector calculus

The operator grad(P) acting upon a scalar P, such as pressure or temperature, and the operators div($\tilde{v}$) and curl($\tilde{v}$) acting upon the vector $\tilde{v}$, which could represent the velocity, were created to describe the flow of liquids. They represent the *gradient* of the scalar P, the *divergence* of a fluid flow that is diverging and the *curl* or rotational motion of a fluid flow. We do not rely upon vector calculus to derive any result in this book, nor is it necessary to fully understand the properties of such vector operators. We will simply use these vector operators as a convenient shorthand in order to write compactly expressions which, when written in full in terms of their components, are lengthy.

The **gradient** of a scalar number P is written as grad(P) or as $\tilde{\nabla}P$ and is a vector $\tilde{v}$ with components given below. These components measure the variation of the scalar quantity P along different directions which we chose to be the x-, y- and z-axes of Cartesian coordinates:

$$\tilde{v} = \text{grad}(P) = (v_x, v_y, v_z) = \left(\frac{dP}{dx}, \frac{dP}{dy}, \frac{dP}{dz}\right).$$

The **divergence** of a vector $\tilde{v}$ is a scalar number N and is written as div($\tilde{v}$) or as $\tilde{\nabla} \cdot \tilde{v}$, like a scalar product, such that

$$N = \text{div}(\tilde{v}) = \frac{dv_x}{dx} + \frac{dv_y}{dy} + \frac{dv_z}{dz}.$$

The divergence is seen as a scalar measure of the sum of changes of the components with the distances along the directions of the components.

The **curl** of a vector $\tilde{v}$ is itself a vector $\tilde{C}$, written as curl($\tilde{v}$) or as $\tilde{\nabla} \times \tilde{v}$, as in a vector product; it may be written in terms of its components as

$$\text{curl}(\tilde{v}) = \tilde{C} = (C_x, C_y, C_z),$$

where

$$C_x = \frac{dv_z}{dy} - \frac{dv_y}{dz}, \quad C_y = \frac{dv_x}{dz} - \frac{dv_z}{dx}, \quad C_z = \frac{dv_y}{dx} - \frac{dv_x}{dy}.$$

Note that the forms of the vector operators are easily remembered and are consistent with our definition of the scalar and vector products if the delta operator $\tilde{\nabla}$ is considered as a vector operator with components.

$$\tilde{\nabla} = \left(\frac{d}{dx}, \frac{d}{dy}, \frac{d}{dz}\right).$$

Thus the gradient of the scalar quantity P becomes the vector $\tilde{\nabla}$ multiplied by the scalar P as below.

$$\text{grad}(P) = \tilde{\nabla}P = \left(\frac{d}{dx}, \frac{d}{dy}, \frac{d}{dz}\right)P = \left(\frac{dP}{dx}, \frac{dP}{dy}, \frac{dP}{dz}\right).$$

The divergence of the vector $\tilde{v}$ becomes the scalar product of the vectors $\tilde{\nabla}$ and $\tilde{v}$ so that

$$\text{div}(\tilde{v}) = \tilde{\nabla} \cdot \tilde{v} = \left(\frac{d}{dx}, \frac{d}{dy}, \frac{d}{dz}\right) \cdot (v_x, v_y, v_z) = \frac{dv_x}{dx} + \frac{dv_y}{dy} + \frac{dv_z}{dz}.$$

Finally, the curl of the vector $\tilde{v}$, which is written as the vector product $\tilde{\nabla} \times \tilde{v}$, becomes

$$\text{curl}(\tilde{v}) = \tilde{\nabla} \times \tilde{v} = \left(\frac{d}{dx}, \frac{d}{dy}, \frac{d}{dz}\right) \times (v_x, v_y, v_z)$$

$$= \left(\frac{dv_z}{dy} - \frac{dv_y}{dz}, \frac{dv_x}{dz} - \frac{dv_z}{dx}, \frac{dv_y}{dx} - \frac{dv_x}{dy}\right).$$

A1.6.   Integrals

A **line integral** is a one-dimensional integral which represents a summation of small increments along a given path in three dimensions. Say we wish to compute the length L of a particular path starting at the point a with Cartesian coordinates (x_1, y_1, z_1) and ending at a point b with coordinates (x_2, y_2, z_2). At a particular point on the path with coordinates (x, y, z), let the next increment of length along the path be dR. The length of the path is then determined by the sum of all the increments along the path dR such that line is traversed once. Such a summation may be written as the line integral

$$L = \int_a^b dR.$$

In electrostatics the line integral is most often used to compute the total change in a scalar parameter such as the voltage $V(x, y, z)$ when a specified path is traversed. At a point (x, y, z) on the path, the next incremental change in the voltage $dV(x, y, z)$ along the next increment of the path dR is by definition (see equation [1.8])

$$dV(x, y, z) = -E_R(x, y, z) \cdot dR.$$

Here $E_R(x, y, z)$ is the component of the electric field $\tilde{E}(x, y, z)$ at the point (x, y, z) which points along the direction of the next increment in length dR of the chosen path. The total increase in voltage along the path is then given by the equation.

$$V(b) - V(a) = \int_a^b dV(x, y, z) = -\int_a^b E_R \cdot dR.$$

The line integral around a particular closed loop is printed as a line integral with a circle through the integral sign. Thus the total length of a given closed loop may be computed as

$$L = \oint dR.$$

A **surface integral** is a two-dimensional integral which represents the summation of some quantity over a specified surface. If we wish to compute the area A of a plane rectangle of height b and length a we would naturally use two-dimensional Cartesian coordinates where the x-axis is parallel to the side of length a and the y-axis is parallel to the side of length b. We locate the bottom left-hand corner of the rectangle at the origin of the coordinates. The element of area is given as $dA = dxdy$ (see the discussion of Cartesian coordinates above). The rectangle is then specified by giving the limits of x and y such that $0 \le x \le a$ and $0 \le y \le b$. The area is then computed by summing all the elemental areas dA such that the total rectangular area is covered exactly once. Such a summation may then be represented by the surface integral shown below as a summation over x and y:

$$A = \int_x \int_y dA = \int_{x=0}^{x=a} \int_{y=0}^{y=b} dxdy = \int_{x=0}^{x=a} dx \int_{y=0}^{y=b} dy = ab.$$

Should we wish to compute the area of a plane circle of radius a we would naturally use two-dimensional polar coordinates where the element of area is $dA = RdRd\theta$ (see the discussion of polar coordinates above). The area can be specified as a sum of the elemental areas dA such that these elemental areas cover the area of the circle exactly once. Such a sum may be represented by a surface integral by specifying the limits on the variables R so that $0 \le R \le a$ and θ so that $0 \le \theta \le 2\pi$:

$$A = \int_R \int_\theta dA = \int_{R=0}^{R=a} \int_{\theta=0}^{\theta=2\pi} RdRd\theta$$

$$= \int_{R=0}^{R=a} RdR \int_{\theta=0}^{\theta=2\pi} d\theta = \frac{a^2}{2}(2\pi) = \pi a^2.$$

Notice that this integral is easy to calculate as it becomes the product of an integral only in R and an integral only in θ. In general, the two integrals in a surface integral are not independent and a simple example of this is given in Box 1.2.

A surface integral is often used in electricity and magnetism to measure the outward flux of some quantity through the specified surface. If at a given point (x, y, z) on the surface the electric field is $\tilde{E}(x, y, z)$, then we can specify a small area of the surface dS at that point and calculate the component of the field E_N normal to that surface element. We can also calculate the two components of the vector $\tilde{E}$ that are perpendicular the direction of E_N and thus tangential to the surface element dS. Only the component E_N penetrates the surface. The other components are tangential to that surface element dS and do not cross the surface element. The elemental **flux of the electric field** passing through dS is defined to be $dJ = E_N dS$. Thus the total flux J of the electric field $\tilde{E}$ through the surface S may be computed by adding all the elemental contributions dJ such that the elemental areas dS cover the surface exactly once. This sum is represented by the surface integral

$$J = \iint_S dJ = \iint_S E_N dS.$$

A **volume integral** is a summation over three dimensions and is useful when calculating volumes or the quantity of some substance contained within the specified volume. To calculate the volume V of a rectangular box with unequal sides of lengths a, b and c we would use three-dimensional Cartesian coordinates with the axes pointing along the edges of the box. The box then has a length a along the x-direction, a length b along the y-direction and a length c along the z-direction. We place the box in the octant of space within which x, y and z are positive with a corner of the box at the origin of the coordinates. The elementary volume element is $dV = dx dy dz$. The volume is then given by the summation of these elementary volumes dV such that they make up the total volume V exactly once. Such a sum may be written as the volume integral shown below, in which the ranges of x, y and z are determined by the lengths of the box along these Cartesian directions:

$$V = \int_{x=0}^{x=a} \int_{y=0}^{y=b} \int_{z=0}^{z=c} dV = \int_{x=0}^{x=a} \int_{y=0}^{y=b} \int_{z=0}^{z=c} dx dy dz$$

$$= \int_{x=0}^{x=a} dx \int_{y=0}^{y=b} dy \int_{z=0}^{z=c} dz = abc.$$

Once again this particular volume integral is simple to calculate because it may be written as the product of three integrals each involving one dimension only. In general this is not so and a volume integral may be easy to write down but difficult to calculate.

To calculate the total amount of some quantity such as charge within a specified volume V we need to know the density of the charge $\rho(x, y, z)$ (the amount of charge per unit volume) at every point within the surface that defines the specified volume. Then the amount of charge within the volume element $dxdydz$ becomes $dQ = \rho(x, y, z)dxdydz$ and the total charge Q within the specified volume may be calculated using a volume integral as below, where the limits of x, y and z are determined such that the summation over the elementary volumes exactly encompasses the specified volume once only:

$$Q = \int_x \int_y \int_z dQ = \int_x \int_y \int_z \rho(x, y, z)dxdydz = \int\int\int_V \rho dV.$$

The first two triple integrals are over the Cartesian coordinates, while the third is a more general notation for the summation of ρdV over all the elementary volumes dV that make up the total volume V and can represent any type of coordinates such as three-dimensional spherical polar coordinates.

A1.7.   Geometrical vector theorems

Twice in the text, when dealing with electrically (equation [4.9]) and magnetically (equation [8.8]) polarizable material, we transform our equations using two geometrical vector theorems. These are abstract mathematical theorems of universal applicability which describe certain properties of space in two and three dimensions. We simply list below the two transformations that they predict without proof. The **divergence theorem** converts the volume integral of div($\tilde{v}$), the divergence of any vector $\tilde{v}$ calculated over any specified volume V, to the surface integral of the flux of that vector $\tilde{v}$ outward through the surface S that defines and encloses that volume:

$$\int\int\int_V \operatorname{div}(\tilde{v})dV = \int\int_S v_N dS.$$

Stokes's theorem converts the line integral of any vector $\tilde{v}$ around any closed path to the surface integral which measures the flux of the vector curl($\tilde{v}$) through any surface S that terminates on that path:

$$\oint v_R dR = + \int\int_S [\operatorname{curl}(\tilde{v})]_N dS.$$

The sign in the equation is positive as above, if the flux penetrates the surface bounded by the closed path in the direction of advance of a right-hand thread rotated in the direction that the path is traversed, when calculating the line integral.

Appendix 2: The Boltzmann distribution, entropy and detailed balanace

A2.1. Disorder and the number of available states

It is our everyday experience that in a spontaneous change, disorder increases – for example, buildings over time may be reduced to a pile of bricks but a pile of bricks never organizes itself spontaneously into a building. In a simpler example, if we scatter a number of coins in a box and subject the box to random shaking we are surprised, on re-examining the box, to find all the coins distributed in a regular pattern. However, if we arrange the coins in an ordered array initially we are not surprised to find after the random shaking that the coins appear to be randomly distributed on the bottom of the box. This has nothing to do with energy as all the configurations of the coins on the bottom of the box have approximately the same energy.

To quantify this effect, we consider a box containing N molecules of a gas at a fixed temperature. We can then enquire how many of the molecules n_L are to be found in the left half of the box and how many n_R are to be found in the right half. We assume that the gas molecules are distinguishable and that each molecule is as likely to be in one half as in the other. Now we can calculate mathematically the number of different ways, $M(N, n_L)$, that N distinguishable particles can be arranged such that n_L molecules are in one half and thus $n_R = N - n_L$ are in the other. We find

$$M(N, n_L) = \frac{N!}{n_L! \, n_R!} = \frac{N!}{n_L!(N - n_L)!},$$

where $a! = a \times (a - 1) \times (a - 2) \times \ldots \times 2 \times 1$. The number n_L is sufficient to define a given 'macrostate' of the system, in this case how many are in each half. The number $M(N, n_L)$ is called the number of 'available states' or the number of distinguishable 'microstates' that correspond to a given 'macrostate'. In Fig. A2.1 we plot the variation of $M(N, n_L)$ with n_L for $N = 50$.

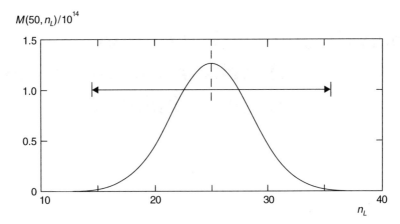

Fig. A2.1   Fifty identical gas molecules are placed in a box. The graph shows the number of ways $M(50, n_L)$ of arranging the molecules such that exactly n_L are to be found within one chosen half of the box, plotted against n_L. The ordinate has been divided by 10^{14} to avoid large numbers. The probability of finding a single molecule in any one half of the box is assumed to be 1/2.

If we associate $M(N, n_L)$, the number of different ways that a macrostate with a given n_L can be formed, with the probability of finding that macrostate, then the most probable state is that with half of the molecules in each half of the box and the least probable states are those with all the molecules within one half. The width of the distribution is characterized by the standard deviation σ which is defined to be the square root of the average value of $(n_L - n_{ave})^2$ summed over all available states. The number n_{ave} is the average value of n_L, which in this case is $N/2$. Probability theory then shows that $\sigma = \sqrt{a \cdot (1 - a) \cdot N}$, where a is the intrinsic probability of each molecule being in the left-hand half of the box. In our case $a = 1/2$, so that $\sigma = \sqrt{N/4}$. The theory goes on to predict that 99.7% of the total number of available states specified by n_L fall within the range $-3\sigma \leq (n_L - n_{ave}) \leq +3\sigma$. For the case $N = 50$ illustrated in Fig. A2.1, $\sigma = 3.54$ and the range $n_{ave} - 3\sigma$ to $n_{ave} + 3\sigma$ is shown in the figure as a double headed arrow between vertical bars.

The standard deviation σ is proportional to the square root of N, while the value of n_{ave} is proportional to N, so that as the number N grows, the size of σ relative to n_{ave} decreases sharply. The graph of the probability of finding a state with a given value of n_L becomes an increasingly narrow peak around the value of n_{ave}. For a value of $N = 10^{23}$, typical of the number of molecules in a laboratory specimen of a real gas, $n_{ave} = 0.5 \times 10^{22}$ and $\sigma = 1.58 \times 10^{11}$. The number n_L which specifies a given configuration of the molecules of the gas has possible values between 0 and 10^{23}, but 99.7% of the total number of available states are to be found in the range between $n_{ave} - 4.74 \times 10^{11}$ and $n_{ave} + 4.74 \times 10^{11}$. Thus 99.7% of the number of available states are to be found within a range for n_L, centred on n_{ave}, that has a width of less than $N/10^{11}$. The probability peak has become very narrow indeed!

Thus from this viewpoint, all available states are possible but the probability of finding the system with a given value of n_L is proportional to the number of available states associated with this macrostate, namely $M(N, n_L)$. Whereas for the two macrostates with all the molecules in one half of the box the number of available states is only 1, for states with n_L very close to n_{ave}, the number of available states becomes gigantic for large N. The probability of finding any state with n_L substantially different from n_{ave} becomes negligible. The macrostate with all the molecules in one half of the box will occur but its probability is so low that we would have to wait on average a time much longer than the age of the universe to observe it.

A2.2.   The Boltzmann distribution

Ludwig Boltzmann extended this theory to the case in which N distinguishable particles can each have a set of energies U_i. If n_i is the number of particles in the state with energy U_i, then the number of available states associated with the macrostate specified by a given set of numbers n_i is given by $M(N, n_1, n_2, \ldots, n_i, \ldots)$ in the equation

$$M(N, n_1, n_2, \ldots n_i, \ldots) = \frac{N!}{(n_1! n_2! \ldots n_i! \ldots)}.$$

Boltzmann then calculated the probability $P(U_i)$ that a given particle was in the state with energy U_i, making three assumptions:

1. The number of particles is fixed.
2. The total energy of the system is fixed.
3. The equilibrium state of the system is the state with the maximum associated value of $M(N, n_1, n_2, \ldots, n_i, \ldots)$.

He found that $P(U_i)$ is given by

$$P(U_i) = \frac{\exp\left(-\frac{U_i}{kT}\right)}{Z} \quad \text{where} \quad Z = \sum_i \exp\left(-\frac{U_i}{kT}\right).$$

The function Z is known as the partition function, $k = 1.38 \times 10^{-23}$ is the Boltzmann constant, and T is the absolute temperature. This Boltzmann distribution is often quoted in the form of a ratio of the probabilities of finding the particle in states of energy U_i and U_j:

$$\frac{P(U_i)}{P(U_j)} = \frac{\exp\left(-\frac{U_i}{kT}\right)}{\exp\left(-\frac{U_j}{kT}\right)} = \exp\left(-\frac{U_i - U_j}{kT}\right).$$

If we have an assembly of a large number of identical systems each with the same set of energy levels with energies U_i all in equilibrium at temperature T, then we can assume that the ratio of the number n_i of systems in the state of energy U_i to the number n_j of systems in the state of energy U_j is equal to the ratio of the probabilities as in the expression

$$\frac{n_i}{n_j} = \frac{P(U_i)}{P(U_j)} = \exp\left(-\frac{U_i - U_j}{kT}\right).$$

We use this form when discussing the cell plasma membrane as an effective barrier to the passage of small ions in Chapter 5 and in the treatment of the Debye layer in Chapter 12.

A2.3.   Entropy

Another important contribution by Boltzmann was to quantify our intuitive concept of disorder. In a system with M available states he suggested that we measure the entropy of that system using a function S where

$$S = k \log_e(M).$$

We believe that all assemblies will spontaneously evolve such as to maximize the total entropy of the assembly and its surroundings. The eventual steady state reached by an isolated assembly is that with maximum entropy. The entropy S is a better measure of disorder than is M itself. If we were to combine two identical systems in the same state, each with M available states, the number of available states of the combined system is M^2, but the entropy S only increases by a factor of 2, in tune with our intuition.

A2.4.   Detailed balance

The Boltzmann distribution applies to a system in equilibrium. Many biological systems are far from equilibrium so that we cannot apply the Boltzmann distribution directly, but we can sometimes gain valuable information about the transition probabilities within a system that is not in equilibrium using the Boltzmann distribution. First we must define a 'temperature bath' to characterize the environment of the particular systems we wish to study. Often we are particularly interested in some subsystem of the whole assembly. For example, we may be interested in the energy states of one particular type of molecule in aqueous solution. However, the vast bulk of the thermal energy of the whole assembly resides in the mechanical translational, vibrational and rotational motions of all the molecules in the solution, dominated by the water molecules. We then say that the particular set of molecules forms the system of interest but that this system is in intimate thermal contact with a bath at temperature T, where the 'bath' consists of the tightly coupled assembly of other molecules in solution. If, for example, we shine light on the molecules of our system so as to induce transitions between their energy levels, then the resultant average occupation numbers of the various energy states within the molecules may not correspond to equilibrium at temperature T. However if the energy input from the light is small in comparison with the thermal energy stored in the bath, then the bath will still be in equilibrium at temperature T. It is an important characteristic of a 'bath' that it has sufficient heat capacity to exchange energy with the system of particular interest without appreciably changing its temperature. It is in these circumstances that we can obtain information about a non-equilibrium system using Boltzmann theory.

Figure A2.2 shows the energy levels available to each of the particles composing some system of interest and in particular two energy levels with energies U_i and U_j. We define $Tr(i, j)$ as the transition probability of a particle initially occupying the state of energy U_i transferring to that of energy U_j, and $Tr(j, i)$ as the transition probability of the reverse transition. This means that if at some time the number of particles occupying the state of energy U_i is n_i, then the number of particles making a transition from the state with energy U_i to the state with energy U_j in unit time is $n_i Tr(i, j)$. If a molecule in our system makes a thermally activated upward energy transition between

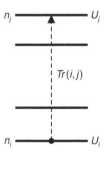

Fig. A2.2   The energy levels U_i available to a particular set of particles together with the numbers of particles n_i which occupy a given energy level. The transition probability per unit time that a particle will make the transition from the energy state specified by i to another specified by j is shown as $Tr(i, j)$.

states of energy U_i and U_j this requires the transfer of energy $U_j - U_i$ to the molecule from the bath, and the reverse transition within our system requires the reverse transfer. Thus the probability of making thermally activated transitions of a given energy within the system is related to the probability of finding transitions of the same energy in the opposite direction in the bath.

Let us first assume that the system is undisturbed and is in a steady state at equilibrium at temperature T with the number of particles occupying the state with energy U_i given by n_i. These average occupation numbers must remain constant in time in the equilibrium state so that we can further assume a **detailed balance** in the upward and downward transitions between any two energy states. Then we must have

$$n_i Tr(i,j) = n_j Tr(j,i).$$

Using the equilibrium Boltzmann ratio for the ratio of the occupation numbers n_i and n_j, we may write

$$\frac{Tr(j,i)}{Tr(i,j)} = \frac{n_i}{n_j} = \exp\left(-\frac{U_i - U_j}{kT}\right).$$

Essentially because the transition probabilities for the particles that form the system are determined by the occupation numbers of states within the equilibrium bath, the relationship between the ratio of transition probabilities and the energies of the states in the system holds **even when the system is not in equilibrium**. Thus when the system is in equilibrium, both the ratio of the occupation numbers and the ratio of the transition probabilities are related to the energies of the system energy levels through the Boltzmann relation written above. When the system is not in equilibrium but is in thermal contact with a bath at temperature T, then only the ratio of the transition probabilities is related to the energies of the energy levels:

$$\frac{Tr(j,i)}{Tr(i,j)} = \exp\left(-\frac{U_i - U_j}{kT}\right)$$

This relationship is important in computer simulations of non-equilibrium biological systems when the kinetics of a non-equilibrium system may be modelled by knowing the ratios of the transition probabilities. An example of such modelling is discussed in Chapter 15.

A2.5.   An entropic force

As an example of a force that has its origins in the tendency of any system to maximize its entropy, we consider a chain that consists of N weightless links that is in contact with a temperature bath at temperature T. We assume that each link can only point to the left or the right, so that the angle between successive links is either $0°$ or $180°$. If the links each have length d and there are n_L links pointing to the left and n_R links pointing to the right, then the length L of the chain is $(n_L - n_R) \cdot d$ provided that $n_L \geq n_R$. We will assume that there are no forces acting between the freely pivoted links so that the probability of a given link pointing left or right is 1/2.

In exact analogy with the problem of the N gas molecules considered in the first paragraph, the number of microstates $M(N, n_L)$ of the chain that corresponds to the macrostate defined by the number n_L is

$$M(N, n_L) = \frac{N!}{n_L!(N - n_L)!}.$$

The most probable state is that with $n_L = N/2$ and for large N the distribution will be sharply peaked about this value. Extended states of the chain with values of n_L very different from $N/2$ will be rare. If we apply equal-amplitude but oppositely directed forces to the chain ends, tending to extend the chain, these will be opposed by the tendency of the chain to adopt more probable configurations corresponding to larger values of $M(N, n_L)$ with n_L closer to the most probable value of $N/2$. In terms of the entropy, we describe this situation by saying that the tension in the chain is created by the tendency of the chain to adopt configurations of greater entropy $S(N, n_L)$, where we have the relation

$$S(N, n_L) = k \log_e[M(N, n_L)].$$

Such forces are sometimes called **entropic forces** because they have their origin in the tendency of the system to be found in the most probable states with the highest entropy. It is true that we have assumed that there are no forces between the links that could contribute to the measured tension in the chain, but the physical origin of the tension lies in the forces exerted by the molecules in the bath which collide with the chain and so flip the links from left to right and vice versa. In Appendix 3 we will calculate the tension in such a chain at a temperature T using thermodynamic arguments.

Appendix 3: An introduction to thermodynamics and the chemical potential

A3.1. The first law

Thermodynamics is a large subject of universal application and we can here only touch on some of its predictions that are directly applicable to material dealt with in this book. In equilibrium thermodynamics we often deal with a system which consists of a collection of particles of particular interest and a temperature 'bath'. The temperature bath exists at a particular temperature T and has a very large thermal capacity so that it can exchange heat energy with the system and yet maintain its temperature constant. The system and bath are discussed from a statistical mechanics point of view in Appendix 2. If the system can exchange heat but not particles with the bath it is called a **closed system**. If it can exchange both heat and particles with the bath it is called an **open system**.

The first law of thermodynamics is a statement of the conservation of energy and a classification of different kinds of energy. For a closed system the first law of thermodynamics states that

$$DQ = dU + pdv$$

where DQ is an increment of heat added to the system, dU is the subsequent incremental rise in the internal energy of the system, p is the pressure and v is the volume of the system so that $p \cdot dv$ is the mechanical work **done by** the system in expanding its volume by a small amount dv. The internal energy encompasses all the internal modes of the system capable of absorbing energy such as the vibrations and rotations of its molecules. The increment in the internal energy U is written with a small d as dU, which indicates that the internal energy is a single-valued function of the parameters such as T and p that specify the state. Mathematically dU is called a **perfect differential** and U is called a function of state. Such a function has the property that

$$\oint dU = 0,$$

meaning that the total integrated change in U, resulting from any series of changes that bring the system back to its initial state, is zero. In this respect U is like the voltage V in electrostatics discussed in Section 1.6. The increment in heat energy DQ is known as an **imperfect differential**. It is written with a capital D, indicating that the heat energy Q is not a function of state and is not a single-valued function of the parameters that specify the state. Thus knowing the parameters that specify the system will not enable us to predict the heat content of the system. To do so we need to know the complete history of the system since it was formed. It is an aim of thermodynamics to replace parameters that are not functions of state with those that are.

For an open system the first law of thermodynamics is written

$$DQ = dU + pdv - \sum_i \mu_i dn_i.$$

The extra term measures the total work **done on** the system when introducing a small change dn_i in the number of particles of type i at a position within the system where the **electrochemical potential** of that type of particle is μ_i summed over all types of particle. The extra term is negative because it measures work **done on** the system, while the mechanical work term is positive because it measures work **done by** the system. For a particle bearing a charge q at a position where the voltage is V and where the concentration of that type of particle is C_i, the electrochemical potential is given by

$$\mu_i = \mu_{0i}(T,p) + qV + kT \cdot \log_e\left(\frac{C_i}{1}\right).$$

The parameter $\mu_{0i}(T,p)$ is the electrochemical potential of the ith type of particle in a 'standard state' when the voltage is zero and the concentration C_i is 1 in the units used, as then $\log_e(1) = 0$ so that $\mu_i = \mu_{0i}(T,p)$. The term qV is the electrical work done on moving a particle with charge q from a position in which the voltage is zero to a position where the voltage is V. The last term, which includes the Boltzmann constant k and the absolute temperature T, measures the work done in changing the local concentration of that type of particle from the concentration specified in the standard state to a concentration of C_i. The term inside the logarithm is written as $C_i/1$ rather than simply as C_i as it must be a dimensionless number expressing the concentration as a fraction of the concentration in the 'standard state' of 1. The definition of the electrochemical potential when applied to ionic solutions should strictly be expressed in terms of ionic activities rather than ionic concentrations. However, for simplicity we will always assume that the solutions are sufficiently dilute to approximate to ideal solutions so that we can use concentrations rather than activities.

A3.2.   The second law

The **second law** of thermodynamics states that in a **reversible change** the increase in the heat DQ is given by the expression below, where S is a function of state called the **entropy**

$$DQ = TdS \quad \text{or} \quad dS = \frac{DQ}{T}.$$

Note that this equation allows us to replace an imperfect differential DQ with an expression $T \cdot dS$ which contains only perfect differentials. In an irreversible change the increase in entropy dS is greater than dQ/T. However, in most biological systems the temperature and pressure may be considered as constant, and all heat transfers at constant temperature are reversible. When dealing with biological systems at constant temperature we may assume the reversible form of the second law of thermodynamics. The entropy measures the disorder in the system, or, more accurately, the number of available states, as discussed in Appendix 2. We believe that all systems will evolve spontaneously such as to maximize their entropy if left undisturbed.

For a closed system the combined first and second laws yield the equation

$$dU = TdS - pdv.$$

For an open system the equation becomes

$$dU = TdS - pdv + \sum_i \mu_i dn_i.$$

A3.3.   The Gibbs function

The Gibbs function, G, is a function of state and dG is a perfect differential. It is defined by

$$G = U - TS + pv.$$

If we differentiate this expression we obtain

$$dG = dU - TdS - SdT + pdv + vdp.$$

To determine the change $dG_{T,p}$ in the function G at constant T and p, we substitute $dT = dp = 0$ so that

$$dG_{T,p} = dU - TdS + pdv.$$

We will now consider a closed system for simplicity which can exchange heat but not particles with a bath at temperature T. Let us suppose that a small increment of heat DQ is transferred from the bath at temperature T to the system also at temperature T, as sketched in Fig. A3.1. Substituting the value of $dU - TdS = DQ$ from the first law into our expression for $dG_{T,p}$, we obtain the equation

$$dG_{T,p} = DQ - TdS,$$

where dS measures the increase in entropy of the system. But the transfer of the heat DQ from the bath to the system has, according to the second law, diminished the entropy of the bath S_B so that

$$dS_B = -\frac{DQ}{T} \quad \text{or} \quad DQ = -TdS_B.$$

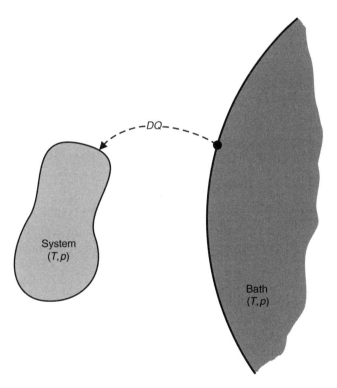

Fig. A3.1   A diagram illustrating the transfer of an increment of heat DQ from a temperature bath at temperature T and pressure p to the system studied which is at the same temperature and pressure.

Substituting this value of DQ in the equation above, we obtain

$$dG_{T,p} = -T(dS + dS_B),$$

where S_B is the entropy of the bath and S is the entropy of the system studied.

Algebraically this equation states that a drop in the function $G_{T,p}$ corresponds to a rise in the function $(S + S_B)$ so that to maximize the function $(S + S_B)$ we must minimize the function $G_{T,p}$. Thus we arrive at the important result that if we wish to find **the state which maximizes the entropy of both the system and the bath we need only minimize the Gibbs function for the system alone**. But the state which maximizes the entropy of the system and its surroundings is the equilibrium state, as discussed in Appendix 2, towards which all systems will spontaneously evolve. Thus all spontaneous thermally activated changes will reduce the Gibbs function of the system and the equilibrium state of an undisturbed system will have a minimum value of $G_{T,p}$, so that

$$G_{T,p} \text{ at equilibrium is a minimum} \quad \text{or} \quad dG_{T,p} = 0 \text{ at equilibrium.}$$

Any change of the system that results in a change in $G_{T,p}$ such that $dG_{T,p} \leq 0$ can happen spontaneously without outside intervention but if, as a result of the change, $dG_{T,p} \geq 0$ then this change requires an outside energy source. We can discriminate in this manner between biological changes that can happen spontaneously, such as the passage of ions through passive membrane ion channels, and those requiring an external energy source, such as the active transfer of ions across a membrane performed by some ion pumps.

A3.4. Uses of the chemical potential

Substituting the value of DQ from the first law for an open system into our expression for $dG_{T,p}$, and using the fact that $DQ = T \cdot dS$, we obtain the equation for changes in the Gibbs function when the temperature and pressure are held constant:

$$dG_{T,p} = \sum_i \mu_i dn_i.$$

Because $dG_{T,p}$ and dn_i are perfect differentials it can be shown mathematically that a consequence of the equation above is that

$$\mu_i = \left(\frac{\partial G}{\partial n_i} \right)_{T, p, n_j \neq n_i}.$$

This equation asserts that the chemical potential μ_i is the partial Gibbs function for each particle of type i. The subscripts outside the bracket represent the fact that the differentiation is to be performed under conditions that T and p remain constant, as do all numbers n_j such that $n_j \neq n_i$. In words, the chemical potential for a particle of type i is given by the rate of change of $G_{T,p}$ with n_i while all other parameters are kept constant.

If we consider an assembly of particles at equilibrium so that $dG_{T,p} = 0$, and remove a small number dn of particles from a region where the electrochemical potential is μ_i and move them to a position where it is μ_j, then, from the equation above for $dG_{T,p}$ we obtain the equation

$$0 = dG_{T,p} = -\mu_i dn + \mu_j dn \quad \text{or} \quad \mu_i = \mu_j.$$

This demonstrates that within a system in equilibrium the electrochemical potential of a given type of particle is everywhere constant.

As an application of this result we consider a narrow pore penetrating the outer plasma membrane of a cell across which there exists both a voltage difference $V_{in} - V_{out}$ and a concentration difference $C_{in} - C_{out}$. We will further assume that the pore will allow the passage only of cations of type i that have a charge q. We know from Fick's law (see Section 3.3) that molecules spontaneously flow down a concentration gradient and we know that a cation will spontaneously move down a voltage gradient because of the electric field present. But what will happen if there exists simultaneously both a concentration difference and a voltage difference across the membrane each encouraging ion flow in opposite directions? From the equation above we simply require that in equilibrium the electrochemical potential for the cation must be the same at each end of the pore. Thus in the equilibrium steady state we must have

$$qV_{in} + kT \cdot \log_e\left(\frac{C_{in}}{1}\right) = qV_{out} + kT \cdot \log_e\left(\frac{C_{out}}{1}\right)$$

or

$$V_{in} - V_{out} = \left(\frac{kT}{q}\right) \log_e\left(\frac{C_{out}}{C_{in}}\right).$$

The voltage difference $V_{in} - V_{out}$ that ensures equilibrium and hence no net flow through the trans-membrane pore for this type of cation when there exists a concentration ratio (C_{out}/C_{in}), is known as the **Nernst potential**. A positive value of $V_{in} - V_{out}$ will tend to drive cations through the pore outward across the membrane, and a positive value for (C_{out}/C_{in}) will tend to drive the cations inward across the membrane, as deduced from Fick's law. The two tendencies are balanced such that no net flow of cations occurs when the equation above linking these quantities is obeyed.

Using the fact that if

$$\log_e(y) = x \quad \text{then} \quad y = \exp(x),$$

we can transform the above equation so that it becomes

$$\frac{C_{out}}{C_{in}} = \exp\left[\frac{q(V_{in} - V_{out})}{kT}\right] = \exp\left[-\frac{q(V_{out} - V_{in})}{kT}\right].$$

This is just the prediction of the Boltzmann distribution discussed in Appendix 2, and it is applicable in this case because the system is in equilibrium. In words, it states that the ratio of the probabilities of finding a cation just outside the pore to the probability of finding a cation just inside the pore, which is given by C_{out}/C_{in}, may be written as

$$\frac{C_{out}}{C_{in}} = \exp\left[\frac{(U_{out} - U_{in})}{kT}\right] = \exp\left[-\frac{q(V_{out} - V_{in})}{kT}\right],$$

where $U_{out} = qV_{out}$ and $U_{in} = qV_{in}$ are the electrical energies of the cation just outside and just inside the pore.

Another use of the electrochemical potential is to deduce the forces that act when μ_i varies with position using a technique called the **principle of virtual work**. We allow a particle of type i to move a small distance dx in the x-direction. The move is assumed so small as to leave the rest of the system unchanged. If the prevailing force in the x-direction acting on the particle is written as F_x then, in the move, the force does work $F_x \cdot dx$ (see Appendix 1) so that its electrochemical potential falls. The change in this potential of the particle when it moves a distance dx is $(d\mu_i/dx) \cdot dx$, where $(d\mu_i/dx)$ is the local gradient of the electrochemical potential in the x-direction. Thus equating the work done by the force $F_x \cdot dx$ to the negative change in the electrochemical potential, we obtain

$$F_x \cdot dx = -\left(\frac{d\mu_i}{dx}\right) \cdot dx \quad \text{or} \quad F_x = -\frac{d\mu_i}{dx}.$$

Thus the negative gradient of the electrochemical gradient for a given type of particle in a particular direction gives the prevailing force on that type of particle in that direction. This result is referred to in Section 3.5 when discussing diffusion. Note the similarity in the force generated by a negative gradient of the electrochemical potential and the electric field, which is the force on a unit charge, generated by a negative gradient of the voltage.

A.3.5.   The tension in an 'entropic chain'

Here we will apply our knowledge of thermodynamics to the calculation of the tension F generated in the chain discussed in Section A2.5. We apply equal and opposite forces of F to the ends of the chain of length L which is in equilibrium with a bath of temperature T. The chain is a closed system, and application of the first and second laws of thermodynamics to the chain results in the equation

$$dU = TdS + FdL.$$

To obtain this equation we have assumed that the volume v of the chain does not change when it is extended so that $dv = 0$. The term representing the work done on the chain by the externally applied forces, FdL, appears with the opposite sign to the work term pdv for a gas because FdL represents the work **done on** the chain and the term pdv represents the work **done by** the gas. In this system there are no forces between the links so that $dU = 0$ and so $FdL = -TdS$. Using the mathematical result that

$$\left(\frac{dW}{dx}\right)_Y = \left(\frac{dW}{dz}\right)_Y \left(\frac{dz}{dx}\right),$$

we arrive at the result that at constant temperature

$$F = -T\left[\frac{dS}{dL}\right]_T = -T\left[\frac{dS}{dn_L}\right]_T \left[\frac{dn_L}{dL}\right]_T.$$

Now $L = (n_L - n_R)d = (2n_L - N)d$, where N is the total number of links each of length d. Thus we have the equations

$$\frac{dL}{dn_L} = 2d \quad \text{and} \quad F = -\left(\frac{T}{2d}\right)\left[\frac{dS}{dn_L}\right]_T.$$

The force F is positive because, for the case $n_L \geq N/2$ we are considering, $M(N, n_L)$ and hence $S = k\log_e[M(N, n_L)]$ fall in value as n_L increases. The force is proportional to the temperature T because the greater the temperature the greater the tendency of the chain to fold by interaction with the bath at temperature T. This model chain can be considered as a very much simplified model of polymer chain folding or of the tension exerted by substances such as rubber.

The differentiation of the full expression for the entropy is not straightforward. However, if N, n_L and n_R are each large numbers we can use Stirling's approximation for a large number N,

$$\log_e(N!) \cong N\log_e N - N;$$

then we may write

$$S = k \log_e \left[\frac{N!}{n_L! n_R!} \right] = k[\log_e(N!) - \log_e(n_L!) - \log_e(n_R!)]$$

$$\cong k\{N \log_e N - n_L \log_e n_L - n_R \log_e n_R\},$$

where we have used the fact that $N = n_L + n_R$. Thus finally, because $dn_L = -dn_R$, we arrive at a simple approximate expression for the tension F in the chain:

$$F = -\left(\frac{T}{2d}\right)\left(\frac{dS}{dn_L}\right)_T \cong +\left(\frac{kT}{2d}\right) \log_e \left(\frac{n_L}{n_R}\right) = \left(\frac{kT}{2d}\right) \log_e \left[\frac{n_L}{(N - n_L)}\right].$$

To obtain this expression we have used the mathematical results that

$$\text{if} \quad A = n \log_e(n) \quad \text{then} \quad dA = dn \log_e(n) + n\left(\frac{dn}{n}\right) = dn[\log_e(n) + 1],$$

and that

$$-dn_L \log_e(n_L) - dn_R \log(n_R) = -dn_L \log_e(n_L) + dn_L \log_e(n_R)$$

$$= -dn_L \log_e \left(\frac{n_L}{n_R}\right).$$

Appendix 4: Hints for the solution of and numerical answers to the problems

Chapter 1

1.1   An exercise in the relationship between electric field and voltage discussed in Section 1.6.

1.2   Apply Gauss's law to circular cylinders of radius R that have the wire as axis to find the radial component. The component parallel to the wire is obtained by invoking the symmetry of the long wire.

1.3   Note that the y-components of $\tilde{E}$ cancel.

(a) $E_x(x) = 2 \left[\dfrac{Q}{4\pi\varepsilon_0(a^2 + x^2)} \right] \left[\dfrac{x}{(a^2 + x^2)^{1/2}} \right]$

(b) $V(x) = \dfrac{2 \cdot Q}{4\pi\varepsilon_0(a^2 + x^2)^{1/2}}$ and $E_x = -\dfrac{dV(x)}{dx}$

1.4 $Q = 9.64 \times 10^{-2}\,\mathrm{C}$. Force $= 8.35 \times 10^9\,\mathrm{N} =$ weight of $8.51 \times 10^8\,\mathrm{kg}$.

1.5   Worked example.

1.6   A direct application of the result derived in Problem 1.5.

1.8   Apply Gauss's law to spherical surfaces of radius R centred on the sphere. For the case $R < a$, use the fact that the charge inside a sphere of radius R ($R < a$) is given by $\frac{4}{3}\pi R^3 \rho$.

1.9   Worked example.

1.10   To confine a positive charge at a given point, the voltage must rise for an excursion in any direction. Show, using the result derived in Problem 1.7, that this is not possible.

Chapter 2

2.1   Distribute the charge Q symmetrically on the top and bottom faces of the centre plate and apply the techniques used in Section 2.4 to obtain

$$V = \frac{Qd}{2A\varepsilon_0}, \quad C = \frac{2A\varepsilon_0}{d}.$$

2.2   An exercise in capacitances.

2.3   Calculate $E(R)$ using Gauss's law applied to spheres of radius R co-centric with the shell for $R > a$ and $R < a$. Use equation [1.11] to deduce $V(R)$ using the fact that $V(\infty) = 0$.

2.4   In which regions of space do electric fields exist in the two possible charge configurations of the sphere and the shell? What capacitances are associated with these fields?

2.5 $\sigma_s = 7 \times 10^{-4} \, \mathrm{C\,m^{-2}}$.

2.6 $E_M = 10^7 \, \mathrm{V\,m^{-1}}$, $U_M = 3.92 \times 10^{-15} \, \mathrm{J}$, $U_Q = 8.0 \times 10^{-21} \, \mathrm{J}$, $U_T = 4.05 \times 10^{-21} \, \mathrm{J}$.

2.7 $708 \, \mu\mathrm{F}$.

Chapter 3

3.1 $I = NqAv_D$. Thus $v_D = 2.35 \times 10^{-5} \, \mathrm{m\,s^{-1}}$.

3.2 $V_1 = \left(\dfrac{\rho_1 L_1}{\rho_1 L_1 + \rho_2 L_2} \right) V, \quad V_2 = \left(\dfrac{\rho_2 L_2}{\rho_1 L_1 + \rho_2 L_2} \right) V$

3.3 $t = 1.25 \times 10^{-2} \, \mathrm{s}, \quad E_x = 1.55 \times 10^5 \, \mathrm{V\,m^{-1}}, \quad \theta = \tan^{-1}\left(\dfrac{1}{10} \right) = 5.7°$

3.4 $\sigma = 0.164 \, \Omega^{-1} \, \mathrm{m^{-1}}$.

3.5   Note that in the expressions for the current density J given in equations [3.10] and [3.11], the concentration is the number of ions per cubic meter and not molar. $J(\mathrm{H^+}) = 3.48 \times 10^{-4} \, \mathrm{A}$, $J(\mathrm{OH^-}) = 1.98 \times 10^{-4} \, \mathrm{A}$, $J(\mathrm{Na^+}) = 0.498 \, \mathrm{A}$ and $J(\mathrm{Cl^-}) = 0.760 \, \mathrm{A}$.

3.6   Acceleration required is $1.42 \times 10^5 \, \mathrm{g}$.

3.7   In the steady state, when there is no charge accumulation, deduce that the flux of the current density $\tilde{J}$ out of any closed surface within the conductor must be zero. Given that $\tilde{J} = \sigma \tilde{E}$ (equation [3.4]), where σ is the constant conductivity of the conductor, this proves that the flux of $\tilde{E}$ out of any closed surface within the conductor is zero or $\mathrm{div}(\tilde{E}) = 0$. Use this fact, together with the fact that $-\oint E_R dR = V$, to solve the problem.

Chapter 4

4.1 $U(R, \theta) = \dfrac{pQ \cos(\theta)}{4\pi\varepsilon_0 R^2}$, $U(R, 0) = +\dfrac{pQ}{4\pi\varepsilon_0 R^2}$, $U(R, 90°) = 0$, $U(R, 180°) = -\dfrac{pQ}{4\pi\varepsilon_0 R^2}$.

4.2   Discuss the stability of the equilibrium.

4.3 $\Delta U = 2pE$, $E = 3.28 \times 10^8 \, \text{V m}^{-1}$

4.4 $d = 0.273 \times 10^{-9}$, $E(1) = 1.93 \times 10^{10} \, \text{V m}^{-1}$, $E(80) = 2.41 \times 10^8 \, \text{V m}^{-1}$, $2pE(1) = 2.38 \times 10^{-19} \, \text{J}$, $2pE(80) = 2.98 \times 10^{-21} \, \text{J}$.

4.5   Worked example.

4.6 $\sigma_B = \varepsilon_0 \left(\dfrac{\varepsilon_{R1} - \varepsilon_{R2}}{\varepsilon_{R2}} \right) E_1$.

4.7   Needle cavity $E_C = E$. Disk cavity $E_C = \varepsilon_R E$. Calculate the excess field in the disk cavity due to the bound surface charges as we did for the parallel plate capacitor in Section 2.4.

Chapter 5

5.1 $E_I = 1.43 \times 10^3 \, \text{V m}^{-1}$, $P_I = 7.59 \times 10^{-8} \, \text{C m}^{-2}$, $\sigma_S = P_N$, $P_{\text{area}} = 7.59 \times 10^{-11} \, \text{C m}^{-1}$

5.2 $U = \dfrac{1}{4\pi\varepsilon_0} \left[\dfrac{4Q_1 Q_2}{a} + \dfrac{Q_1^2 + Q_2^2}{\sqrt{2}a} \right]$

5.3 $\theta_2 = 0$, $68.6°$, $80.3°$, $85.3°$, $88.6°$.

5.4 $Q_B = Q \left(\dfrac{1}{\varepsilon_{R2}} - \dfrac{1}{\varepsilon_{R1}} \right)$. The voltage at the surface of the cavity is

$$V(a) = - \int_{\infty}^{a} E_R dR = - \int_{\infty}^{a} \dfrac{QdR}{4\pi\varepsilon_0\varepsilon_{R2}R^2} = +\dfrac{Q}{4\pi\varepsilon_0\varepsilon_{R2}a}.$$

5.5   Worked example.

5.6   The charge on the capacitor is $Q_{\text{cap}} = CV$. Does C go up or down when the dielectric is inserted?

5.7   The charges on the opposite plates of the two capacitors must be the same, $+Q$ and $-Q$. Show that this must be so, using a diagram of the two capacitors in series. The voltages across the two capacitors add to equal the total voltage between the terminals.

5.8 $C_1 = \dfrac{\varepsilon_0\varepsilon_R A}{t}$ and $C_2 = \dfrac{\varepsilon_0 A}{d - t}$

5.9 $F = \dfrac{1}{2}\dfrac{Q^2}{A\varepsilon_0} = \dfrac{1}{2}QE$

5.10 $C = 5.56 \times 10^{-9}$ C. $Q = CV$. Thus the number of univalent cations is 1.74×10^9.

Chapter 6

6.1   A radial outward force acts on the wire.

6.2   How does the interaction energy $U(x) = -\tilde{m}\tilde{B}(x)$ change with x?

6.3   (a) Interaction energy $U(R) = -m_1 B_2 = -\dfrac{2m_1 m_2 \mu_0}{4\pi R^3}$. How does $U(R)$ vary with R? (b) Calculate the force on the second coil caused by the components of magnetic field created by the first coil, (1) perpendicular to and (2) in the plane of the second coil.

6.4 $B_Z(R, \theta) = B_R(R, \theta)\cos(\theta) - B_\theta(R, \theta)\sin(\theta) = \left(\dfrac{\mu_0 m}{4\pi R^3}\right)(3\cos^2(\theta) - 1)$

6.5   The interaction energy is

$$U = -m_{2R}B_{1R} - m_{2\theta}B_{1\theta} = \left(\dfrac{m_1 m_2 \mu_0}{4\pi R^3}\right)[\sin(\theta_1)\sin(\theta_2) - 2\cos(\theta_1)\cos(\theta_2)]$$

For $\theta_1 = 90°$, $\cos(\theta_1) = 0$, $\sin(\theta_1) = 1$ and $U = \left(\dfrac{m_1 m_2 \mu_0}{4\pi R^3}\right)\sin(\theta_2)$.

6.6   Replace coil by equivalent magnetic dipole $m = I\pi a^2$ to obtain $B = \dfrac{\mu_0 I a^2}{4R^3}$ perpendicular to the plane of the coil.

Chapter 7

7.1 $B_T(R)_{\text{inside}} = \dfrac{\mu_0}{2\pi R}I\left(\dfrac{\pi R^2}{\pi a^2}\right) = \dfrac{\mu_0 I R}{2\pi a^2}$, $B_T(R)_{\text{outside}} = \dfrac{\mu_0 I}{2\pi R}$

7.2 $B(0,0) = \dfrac{\mu_0 I}{\pi a}$ along the negative z-axis. $B(0, a/2) = 2.67\dfrac{\mu_0 I}{2\pi a}$; $B(0, 2a) = \dfrac{\mu_0 I}{3\pi a}$; $B(0, y_1) \to 0$ for $y_1 \gg a$.

7.3 $B_T = \dfrac{\mu_0 I}{2\pi R}$

7.4 $F = \dfrac{\mu_0 I^2}{2\pi d}$. Parallel currents attract.

7.6 $B_H = 31.4\ \mu\text{T}$.

7.8   Add the fields of the two coils as given by equation [7.8]. $\left(\dfrac{4}{5}\right)^{1.5} = 0.7155$.

Chapter 8

8.1 $H_{X1} - H_{X2} = J_Y, H_{Y1} = H_{Y2}$

8.3   See Problem 1.4.

8.4   Paramagnets are attracted to regions of high magnetic field.

8.5   Worked example.

Chapter 9

9.1   Flux $= AB\cos(\theta)$ $\therefore$ emf $= ABN\sin(\theta)\frac{d\theta}{dt} = 2\pi fABN\sin(\theta)$.
Therefore $I_{peak} = \frac{2\pi fABN}{R}$ where $A = \pi a^2$.

9.2 $\dfrac{d}{dt}$(Flux of $\tilde{B}$) $\cong \left(\dfrac{10^{-3} \times 50 \times 10^{-6}}{0.25}\right)$. Thus $V_{peak} \cong 2 \times 10^{-7}$ V. No.

9.3   Find the force on the ring due to the interaction between the magnetic field and the current induced in the ring. Does the force reverse when the field reverses?

9.4   Voltage $= B_Y Lv$

9.5   The magnetic field created by the coil threads through all N turns so that the total flux threading the coil becomes N times the flux through a single turn.

9.6   Apply Gauss's law in $\tilde{D}$ and the divergence theorem (equation [4.9]).

Chapter 10

10.2 $V_{rod} = B_Y Lv$, $V_{meter} = 0$. Explain why.

10.3 $B = \dfrac{E}{v}$ applied along the negative z-axis.

10.5 $v_{initial} = \left(\dfrac{qB_0}{m}\right)R_{initial}$ and $v_{final} = \left[\dfrac{q(B_0 + \Delta B)}{m}\right]R_{final}$.

10.6   Faster rotation in the field caused by a circulating electric field that accelerates the particle which is present while the magnetic field is increasing as deduced by Faraday's law.

Index

Ampère's law
  in a vacuum 108–12
  in polarizable material 133–8
ampullae of Lorenzini 213
animal compasses 210–11
animal magnetic compass, mechanisms
  free radicals 218–22
  magnetic induction 211–5
  magnetite crystals 215–8
available states 263–5

Biot–Savart law 123–5
Boltzmann distribution 187–8, 265–6
boundary conditions
  $\tilde{E}$ and $\tilde{D}$ 75–6
  $\tilde{B}$ and $\tilde{H}$ 139–40

capacitor
  cell membrane 39–40
  parallel plate 37–9
  single conductor 40, 88–90
  spherical 41–2
  with dielectric 86–8, 97–8
cell plasma membrane
  as a barrier to ions 93–4
  as a capacitor 39–40
charge
  bound and free 72
  Coulomb units 4
  source of electrostatic fields 4
conductivity 48
coordinates, Cartesian and polar 63–5,
  251–3
coport and counterport models
  electrical 225–7
  kinetics of electrical model 227–34
  mechanical 224–5
Coulomb law 5
current density, bound and free 136
cyclotron motion 159–60

Debye layer
  basic theory 185–8
  Debye length, definition 191
  Debye length table 192
  electric field variation 188–90
  ionic concentrations 194–5

  reliability of simple theory 195–6
  surface voltage for various solutions (table) 190
  voltage variation 191–3
detailed balance 230, 266–8
diamagnetism 164–6
dielectric constant, relative
  definition 74–5
  water as a dielectric 68–9, 81–2
dielectric material, effect of
  capacitors 86–8
  charge distributions 83
  induced surface and body charges 69–72
  spherical conductor 82–4
dipole, electric
  couple acting in a field 66–7
  definition 61
  electric field of dipole 62
  energy in field 67–8
  moment of water molecule 81–2
  polar components of field 65
  voltage of dipole 63
dipole, magnetic
  couple acting in field 105
  energy in field 105
  magnetostatic potential 107
  of small current loop 102, 104–5
  polar components of field 103
displacement current 150–2
dissociation of water and pH
  definition 178
  effect of electric field 181–2
divergence theorem 72, 262

Earnshaw's theorem 27
electric displacement $\tilde{D}$
  definition 73
  dielectric boundary conditions 75–6
  properties 74–5
electric field $\tilde{E}$
  definition 5–6
  dielectric boundary conditions 75–6
  diffraction at dielectric boundary 85–6
  in a conductor 31
  local field 30
  macroscopic field 30
  near a conductor 34–7
  of ions in narrow aqueous pores 201–4

electric screening 32
electrochemical potential 56, 271, 274–5
electromagnet, simple model 143–4
electromagnetic wave 153
electromotive force or *e.m.f* 145–6
electrostatic energy
    of an assembly of charges 91–2
    of a single conductor 88–90
    stored in electric field 92–3
entropy 266
entropic force 268–9, 277–8

Faraday's law of induction 145
flux
    electric field 13–15
    magnetic field 103
force
    on electric charge 6
    on electric current 100–1
    on electric dipole 66–7
    on magnetic dipole 104–6

Gauss's law, electric
    Gauss's law surfaces 17
    in a vacuum 13–15
    with dielectrics 72–4
Gauss's law, flowing fluids 9–10
Gauss's law, magnetic 103

integrals
    line 19–20, 259–60
    surface 10–13, 260–1
    volume 16, 261–2
inverse square law 5
ionizable residues and pK_a
    definition 179–81
    effect of an electric field 181–2
    effect of dielectric environment 182–5
ions
    cations and anions 29
    in bulk aqueous solution 175–8
    interactions in narrow pores 204–5
    radius of common ions (table) 50
    self-energy in bilayer 93–4
    self-energy in narrow pore 198–201
ions, diffusion
    diffusion constants (table) 50
    drift velocity in an electric field 53–5
    Fick's first law 51–3
    random walk 49

Larmor theorem 161–3
Lenz's law 146, 165
light, velocity 153
lines of force 8–9
Lorentz force 158–9

magnetic field $\tilde{B}$
    boundary conditions 139–40
    diffraction at boundary 140

homogeneity of laboratory fields 127–8
    of Helmholtz pair of coils 126–7
    of long solenoid 121–3
    of macroscopic current loop 106–7
    of plane coil 125–6
    of short solenoid 131
    of straight wire 119–121
    polar components of dipole field 103
magnetic field of the Earth 208–10
magnetic pole 103
magnetic resonance, semi-classical theory
    90° pulse 242
    180° pulse 244
    application on macroscopic scale (MRI)
        248–50
    application on a microscopic scale 246–7
    longitudinal relaxation time 243–4
    magnetogyric ratio 236
    resonance phenomenon 239–40
    rotating frame 237–40
    spin echo 244–6
    transverse relaxation time 243
magnetic vector $\tilde{H}$
    boundary conditions 139–40
    definition 137
    properties 138–9
Maxwell's equations 152
mobility, absolute 53, 56

Nernst potential 275

Ohm's law 46–8

paramagnetism 133
permitivity, electrical
    of vacuum ε_0 4
    relative ε_R 74–5
Poisson's equation 27
polarizability, magnetic
    of vacuum μ_0 103
    relative μ_R 137–8
potentials, electrostatic and
        magnetic 112–5

relaxation times, magnetic resonance 243–4
resistance 46
resistivity 47

screening of electric fields
    induced electric fields 149–50
    static electric fields 32–3
self-energy, electrostatic
    ions in aqueous solution 93, 175–8
    ions in narrow aqueous pores 198–201
    single isolated conductor 88–90
solid angle 110–111
special relativity and magnetism 166–8
Stern layer 187
Stokes's theorem 136, 262
susceptibility, magnetic 143

thermodynamics
  electrochemical potential 56, 271,
    274–5
  entropy 266
  first law 270–1
  Gibbs function 272–4
  second law 272

vectors
  components 7, 255
  curl 135–6, 258–9
  divergence 72, 258–9
  divergence theorem 72, 262
  gradient 258–9

modulus 6, 255
scalar or dot product 257
Stokes's theorem 136, 262
summation of vectors 7, 256
vector or cross product 101, 257
virtual work, principle 276
voltage, definition 22
voltage difference, definition 18

water
  hydrogen bonding 176
  ice structure 177
  ordering in narrow pores 205–7
work done by forces and couples 253–5